TAPPING
THE
WEST

TAPPING
THE
WEST

HOW ALBERTA'S CRAFT
BEER INDUSTRY BUBBLED
OUT OF AN ECONOMY
GONE FLAT

SCOTT
MESSENGER

TOUCHWOOD

Edited by Curtis Gillespie
Proofread by Alison Strobel
Cover design by Sydney Barnes
Interior design by Setareh Ashrafologhalai
Indexing by Janice Logan

LIBRARY AND ARCHIVES CANADA CATALOGUING IN PUBLICATION

Title: Tapping the West : how Alberta's craft beer industry bubbled out of an economy gone flat / Scott Messenger.

Names: Messenger, Scott, 1975- author.

Description: Includes index.

Identifiers: Canadiana (print) 20190207264 | Canadiana (ebook) 20190207272 | ISBN 9781771513203 (softcover) | ISBN 9781771513210 (HTML)

Subjects: LCSH: Beer industry—Alberta—History. | LCSH: Breweries—Alberta—History. | LCSH: Microbreweries—Alberta.

Classification: LCC TP573.C3 M47 2020 | DDC 338.4/766342097123—dc23

TouchWood Editions gratefully acknowledges that the land on which we live and work is within the traditional territories of the Lkwungen (Esquimalt and Songhees), Malahat, Pacheedaht, Scia'new, T'Sou-ke and W̱SÁNEĆ (Pauquachin, Tsartlip, Tsawout, Tseycum) peoples.

We acknowledge the financial support of the Government of Canada through the Canada Book Fund, and the province of British Columbia through the Book Publishing Tax Credit.

This book was produced using FSC®-certified, acid-free papers, processed chlorine free, and printed with soya-based inks.

Printed in Canada at Friesens

24 23 22 21 20   1 2 3 4 5

For Carolyn and Garry,
with gratitude and love

CONTENTS

INTRODUCTION

A journey into the heart of Alberta craft beer

O N A FRIDAY afternoon, late in May of 2019, our car rolled to a stop on the Deerfoot into Calgary from Edmonton. A sense of dread pooled in my stomach. There was so much we needed to see, so many breweries to visit, and, because families with small children (such as mine) do not generally like the idea of one parent taking off for another city to drink beer, whether or not it's for "research," we had just one night to do it. I sat helplessly in the back seat, watching drivers around us jockey for what were essentially parking spots. Since hitting Calgary at rush hour was my fault, I said nothing of my concern to my friends. Guy was driving and Colin was beside him in the passenger seat.

"Let's stop for lunch in Lacombe," I had said a few hours earlier, back when time seemed an abstract concept we'd been measuring out in pints rather than centimetres of pavement travelled. Regret threatened to undermine what was supposed to be a momentous occasion: the beginning of our trip to a collection of breweries now the largest and densest in the Prairies. Scattered throughout neighbourhoods south of downtown Calgary are more than a dozen of them, serving up lager and ale judged to be among Alberta's finest.

It was around noon that we'd detoured from the hectic Queen Elizabeth II Highway for Lacombe's charming, century-old downtown. In the middle of it, we'd found Cilantro and Chive, a modern but rustic restaurant featuring a craft beer list few restaurants in Edmonton or Calgary can compete with. The salmon chowder was hearty even if in need of a dash of salt and a savoury something I could not quite put my finger on; the pint of dubbel, from Eighty-Eight Brewing, was a rich and malty promise of what we hoped to experience later that day in the Belt. But the meal had made us leisurely. After settling up, we asked Rieley Kay, the restaurant's co-owner, where we should go in town for coffee. Sweet Capone's, he told us—their cannolis were outstanding. We ambled down the street past original brick facades (these had replaced wooden ones after a massive fire in 1906) and around the corner as instructed. Inside, I stared at a glass case containing a dozen varieties of the Italian pastry. "What should I eat?" I asked a smiling young woman with red hair pulled back in a severe ponytail. "Salted caramel," she said, without hesitation. Though my friends chose to stick with coffee alone, I took the recommendation and indulged while they sipped. We did not hurry away. Only an idiot rushes through moments meant to be savoured.

Now, on the Deerfoot, I was ready to blame the cannolis. After loitering in the streets of Lacombe, we were stuck in a jam. I checked a map of breweries I'd printed at home, looked at my watch, and juggled the order according to closing times and the minutes we were losing while immobile. A few moments later, as we crept less than a car-length ahead, I did it again, determined that this trip not be ruined. For the first time in my life as a fan of local craft beer, I had the opportunity to visit not just a single brewery but several *in succession*. Sure, there was no shortage of selection where we'd come from. I was spoiled in Edmonton, home to Sherbrooke Liquor, the now legendary retailer with a beer cooler stuffed with more than six hundred Alberta beers. But this was different. This trip was less about the product and the deluge of choice, and more about the places and people behind it—seeing what brewers have built and all the beer lovers who are drawn to it.

Or at least that was the plan.

A black pickup zipped into an empty space ahead of us and hammered the brakes. I checked my watch, and pictured breweries too full to make room for three weary travellers in search of a promised land. One more time, I ran through my list. Then, as if the universe could no longer bear my anxiety, it delivered us. "There's our exit," said Guy, glancing at an off-ramp jammed with vehicles. We still crept along, but an end was in sight and, with it, the prospect of a beginning. My mood brightened. Maybe the cannolis were worth it after all.

Later, once we'd arrived and dumped our bags at our rundown but clean and comfortable hotel, we had a ride-share take us to the first stop on my revised itinerary: Village Brewery.

I had a perhaps outsized sense of excitement about our first stop, one that my friends, unfamiliar with the history of the place, and of beer in the province, didn't quite share. While they were eager for that first pint of fresh-made beer, I felt as if I were taking the first steps on a kind of pilgrim's journey. Village was one of the last big breweries to open before everything changed in Alberta. Before the end of 2013, regulations were such that breweries had to be able to make at least 500,000 litres of beer a year. In my mind, the brewery was a kind of monument to the lengths to which people once went for craft beer, back before the provincial government decreed that small-batch commercial brewing was, well, no longer illegal. To my friends, it just seemed like a great place to grab a beer. They were right in thinking so, of course. But as we walked up the stairs to Village's taproom, it took all the restraint I could muster to keep from being the professor at the party and resist trying to impress upon my friends how important that day in 2013 really was, that it was a defining moment in Alberta craft beer as we know it today. Because, with it, what was then known as the Alberta Gaming and Liquor Commission (AGLC) and the Government of Alberta eliminated the industry's biggest barrier to entry, a seemingly arbitrary and decades-old law that had all but ensured that only the biggest spenders could afford to play. By 2014, anyone with the proper licence could brew as little or as much as they wanted to. The bar had not merely been lowered; it had effectively been tossed aside. All that the AGLC asked was that local products be "fit and safe for human consumption." Having met the old requirement and started up in 2012, Village was among a small group of breweries, including Ribstone Creek in Edgerton, that marked the end of an era. In Calgary, this put it at the top of my list. I was just as eager for a beer as Colin and Guy, but not just any beer to start. I had an agenda.

Post-2013, craft beer in Alberta boomed. The outcome of the change in regulations was like a levee gradually giving way. A trickle issued forth in 2014 and '15, then turned into a steady stream in '16. A gush has flowed ever since. I'd taken notice of it all when that trickle was beginning to grow and a new brewery seemed to be opening almost every week. A flood seemed inevitable. Convinced that Alberta hadn't seen a sector arise with such determination since the oil sands in the late 1960s, I decided to investigate in the only way that seemed appropriate—first-hand experience. From January 1 to December 31 of 2016, I bought no other beer but that made by Alberta brewers. I recorded my findings in a blog called *One Year of Alberta Beer*, which once managed to catch the attention of one of the local daily newspapers. Instead of describing it as an exceptional work of beer writing, the journalist went with "not typically a beer-snob review, but instead reflections on beer and many other things" (I take that now to mean uninformed and given to tangents). I'd set myself a goal of sampling every brewery in the province and figured that, come July, my project would run dry. To my delight, I failed. For reasons of geography, as some breweries then remained intensely local, and out of deference to my liver, I drank my way through just forty-one breweries out of the roughly fifty in operation by the end of 2016.

But the exercise had made me curious. How did this apparent success happen in such a relatively short period of time? Attempts at economic diversification in Alberta aren't uncommon, but successful ones are. Over the years, the province has tried to transfer some of the eggs from the natural resources basket into a homegrown financial sector, a biotechnology "cluster," bitumen upgrading, and more. Today, Alberta is known for none of those things, and mining and energy accounts for more than a quarter of its gross domestic product, just as it has for years.

I wondered, therefore, if the barley upgrading operation that is craft beer might not be able to cut itself at least a sliver of that economic pie chart. In Ontario, home to 30 percent of the country's craft breweries, small-batch beer was responsible for some 7,500 jobs in 2015 and generated $370 million in revenue in 2017. What about Alberta? Part of my investigative work, I felt, surely involved the breweries of Calgary.

We stepped into the Village taproom, mostly empty except for the seats around the bar, and took a table surfaced entirely with the brewery's bottle caps. No one came to ask what we'd have, so I went to the bar and asked the bartender, an older man in a light-coloured Hawaiian shirt, if that's where we ordered. "At most taprooms you order at the bar," he said in a neutral tone, obviously seeing me for the Calgary craft brewery noob that I was. I got a pint of tea saison. Guy and Colin ordered pale ales.

Later, as we were deciding whether or not to have seconds, a young man in a pink western-style shirt, worn untucked, collected our empties and told us of plans to tap a cask later that night. The cask had been made by the very band that would be playing a set in the brewery after the tapping. We decided we'd come back for that. In the meantime, however, we agreed that the itinerary could not accommodate another beer at Village. Our ride-share arrived and we headed off to Eighty-Eight Brewing.

Eighty-Eight was one of the Belt's newest additions, and it was there we began to get a sense of taproom culture as none of us could have imagined it. While the brewery pays homage to one of Calgary's best years, when the city hosted the Winter Olympics, it also speaks to the future of craft beer in the province, a future owned by the young and fearless. Unlike Village, which was quiet, mostly empty, and had a laid-back, modern pub feel, Eighty-Eight was noisy, packed, achingly hip, and pleasingly ironic. With its main floor decorated in a giddy combination of blue and pink, the place struck me as a celebration of craft beer as much as of its city. Just walking through the door was enough to make you forget that Calgary today, three decades after that halcyon year, was a less jubilant place, struggling to dig its way out of the province's worst economic downturn of modern times. The beer of Eighty-Eight was not for crying in; it was far too good to be sullied by tears. I started with a sour, tart and refreshing, followed by a smooth and easy-going white ale. I left my sample of Jump Street—a boisterously hoppy 10.1 percent triple IPA—for last, taking a few sips before asking Colin, who has a notoriously low tolerance for alcohol, if he wanted the rest. Without saying yes or no, he tossed it back like it was apple juice. Our glasses empty, we reluctantly abandoned the illusion that Calgary's good times had never come to an end and headed outside. But the illusion had worked. After our visit, we felt fantastic about everything, including the future of the city and the province. If all else fails, we'd always have '88. But we also felt hungry. Instead of the ride-share, we decided to take a walk. Happily, the Dandy Brewing Company was just a few blocks up the street. Once again, Alberta craft beer was about to surprise us. And it wasn't just about the beer.

I am no foodie. For me, eating is a utilitarian activity. Food enters my body, is converted to energy, work is completed, the extraneous bits are excreted, and the process is repeated ad nauseam. Frankly, it strikes me as almost an inconvenience, a flaw of biology that evolution has yet to address. Then I had the rollmops at the tasting room at Dandy.

We sat down, ordered several beers in tiny glasses and meals soon after, then watched as two chefs made them a few feet away from us in an open kitchen. After our food arrived, I popped one of the circular strips of fish into my mouth. It was trout, rather than the traditional herring, better reflecting Alberta fare, and was pickled and wrapped around a sweet pickle and drizzled with what I think was crème fraîche. They were stunning, a perfect yin and yang of sweet and savoury. I alternated between the rollmops and something called baked potato pasties. Also stunning. It was as if the starchy innards were scooped out of the skin, mixed with the Dandy equivalent of eleven herbs and spices, and returned to a casing that had been crisped to perfection. I offered a bite to Colin, a former chef.

"I have no idea how they did this," he said, impressed.

Guy, who'd paused a renovation project at home for this trip, took a moment to inspect the room's minimalist decor of polished concrete floors, white walls, and exposed brick. "This is what my basement should look like," he said.

We all agreed that Edmonton, if not Guy's basement, could use a place like the Dandy, with this style of decor and the same calibre of food and beer, including citrusy, floral IPAs, a tropical-tinged sour, and even an interesting lager, toasty and crisp. Colin ate more of my baked potato pasty than I'd hoped. Secretly glad for his fish allergy and Guy's veganism, I had the rollmops to myself. For the first time in a long time, dinner seemed to me time well spent. We called for a rideshare and decided to wait outside.

In the parking lot, I consulted the itinerary. I'd been wrestling with whether or not to visit breweries in the neighbourhood of Inglewood. As I looked at the map, I was momentarily distracted by the realization that, where we stood, we were bookended by Alberta beer history. Yes, to the south, Village represented the tail end of the pre-deregulation craft brewers in the province, but in Inglewood there was a vestige of a pioneer in the truest sense. Now shuttered, the Calgary Brewing and Malting Company once operated there, beginning in 1892, as the makers of Calgary Beer. Its founder, A.E. Cross, was a leader among the province's prototypical small-batch brewers. This cluster of craft breweries, however, is not directly a product of that early initiative. Instead, it's arguably the result of one man's reaction to the restrictive brewing environment and industry that Cross, inadvertently or otherwise, helped to establish. That man was Ed McNally, the founder of Big Rock. And it was there, at Big Rock, that the story of Alberta craft beer, and Village and Eighty-Eight and Dandy and every other brewery in the Belt and beyond, truly begins.

1

BEFORE THE FLOOD

*The province's
pioneers of beer*

A LOT OF WORK FOR A DECENT PINT

The story of Ed McNally, founder of Big Rock Brewery, is now the stuff of Alberta legend. We love this sort of narrative in this province: Entrepreneur sees opportunity where others see only impossibility and proves that a vision can be turned into reality. His story speaks to grit, determination, capitalizing on good luck, overcoming adversity, and cultivating a sophisticated palate while at the same time never denying his love for something as simple as a Creamsicle. McNally had that much-touted, Albertan "can-do" attitude when, really, the province was doing relatively little.

Big Rock opened in Calgary in September 1985, when the price of oil was US$28 a barrel and unemployment in Alberta was 9.2 percent. Interest rates had declined from asphyxiating peaks of 20-plus percent to a merely oppressive 11 percent. Optimism would have been in as short supply as disposable income.

"This was not a great time to start a venture of this magnitude," McNally told *Canadian Business Journal* twenty years in. "Our province was flat on its butt. But I kept investigating the option, and not starting my brewery was not an option."

The reason he had ruled out that possibility has origins in a family vacation that his daughter, Shelagh McNally, told me about. The idea had been brewing for a while, but a trip to Hawaii made it seem not just palatable to McNally, but crucial. On the plane to the islands, he read an article about micro-breweries along the American East Coast.

"My father was fascinated with this because during the time of the Depression, the only place that was working and continued to work was the Sick brewery in Lethbridge [where he grew up]," said Shelagh, referring to Fritz Sick's operation, which began producing the original Old Style Pilsner following Prohibition in Alberta, which ended in 1924 following a plebiscite. "People seemed to have money for beer, even in the worst of times."

But it wasn't that article about micro-breweries that sealed the deal for McNally. The family was on the beach when her father was struck by the need for a beer. He headed off to the corner store and returned with a well-known macro-lager.

"He bought this teeny half-can," said Shelagh. "It had all the beautiful humidity dripping off, and it looked picture-perfect. He popped the lid and took a big swig—and he literally spat it out." With that, she said, the matter was settled. He disappeared to make calls. "He began working on the creation of Big Rock."

Back in Calgary, McNally and four other directors sank $1.5 million into the effort. They picked up a warehouse owned at the time by the bank, said Shelagh, at a price that reflected the desperation of the times. They scored discounts on equipment, too, stuff McNally said that no one else was interested in.

"We bought this building, furniture and all, for $170,000," he said in 1986 in *The Globe and Mail*. "We got [our] bottling line from Coca-Cola in Great Falls, Mont., for 10 cents on the dollar of what it was worth. We got our packaging equipment from Carling O'Keefe and we rebuilt it."

Big Rock's beer was more expensive than that of its national competitors but it was cheaper than European imports. At the time he was interviewed by the *Globe*, McNally was expecting the company to be near the break-even point within fifteen months, despite a slow start.

"Competition here is really tough," he said. "We came on the market at the same time as Coors in Calgary and a lot of places wouldn't try our draught."

But word was getting out that Big Rock was doing things differently, which was how McNally tended to do things throughout the already varied career that preceded his explosive epiphany on a Hawaiian beach. Probably the most straightforward and simplest thing he ever did was farm barley, berries, and organic vegetables in Okotoks, where he lived. His first job was as a land surveyor. Later, he ended up as a reporter for the *Lethbridge Herald* in the city where he grew up. "I had one article that made it in the national press," he told *Business Edge News Magazine* in 2003. "A bear ran through the Waterton Lakes hotel beer parlour. The bear came running in, sniffed around for a while and then ran out the back door." This may have represented Alberta's first instance of local beer writing.

But agriculture seemed ultimately to have the greatest pull on McNally. It exerted itself in many forms. He became involved with cattle and once held a sale of European breeds in a hotel in downtown Calgary.

"My wife was in the parking lot down below and leading the cattle up through the kitchen, right into the ballroom," McNally said in an interview with an alumni magazine of the University of British Columbia, where he got a law degree in 1951. "I think we sold about 24 cattle. I think we averaged over $2,000 a head. We went out of there with a pile of money. I thought we were the smartest guys in town."

As a lawyer, though, he wasn't one to tolerate any bullshit. In a brief retrospective of McNally's life, published following his death at eighty-nine years old in August 2014, *GrainsWest* magazine described him as "a tireless opponent of the Canadian Wheat Board's monopoly on grain." On behalf of a group of farmers, he led a lawsuit against the board in the 1970s. For decades, producers had been forced to sell grain, including barley, to the agency rather than directly to maltsters, likely depriving them of profits. It struck the entrepreneurial-minded McNally as needless market regulation and restriction.

That work with the farmers helped postpone any consideration of retirement. This is when McNally had his first dreams of a career in beer, motivated by two realizations. One, he'd learned that the two-row barley produced in Alberta makes for some of the best malt—beer's key ingredient—in the world. Two, most of it went to cattle feed. This struck him as a missed opportunity.

Even with beer, McNally, sixty years old when Big Rock opened, seemed determined to hoe a hard row. Inspired by trips to Europe, including the Brussels World's Fair in 1958, he launched with a series of ales—bitter, porter and traditional—in a market dominated by watery lagers produced by three companies that accounted for well over 90 percent of the market. In October 1985, the month after opening, McNally said that Big Rock sold just thirty-two kegs.

"Unless people had travelled to England or Europe, they hadn't experienced beer like that," said Shelagh. "It was like going from Wonder Bread to an amazing hand-crafted sourdough—it was a totally different thing. So, it wasn't necessarily met with a whole lot of warm and fuzzies."

But in the same month a year later, they sold 320 kegs. (To compare, in the third quarter of 2018, the company moved 60,988 hectolitres, equivalent to roughly 210,300 kegs.) Not only did Big Rock offer an alternative to beer drinkers, it gave some members of the industry hope for renewal and maybe even the chance to recapture an element gone missing from the business of brewing beer, such as heart, fun, humanity, and craftsmanship. Larry Kerwin had just left his role as brewmaster at the Molson plant in Calgary when he was contacted in 1994 by McNally. The plant was shutting down that year, and the Big Rock founder was looking to scoop up talent.

After twenty-five years in the big leagues, beginning with Calgary Brewing and Malting before moving on to Carling and then Molson through the companies' merger in 1989, Kerwin was feeling the Big Beer blues. "It wasn't much fun being involved in it," he said. "It wasn't so much a brewing company; it was just a marketing company that made beer. Everything was done for the bottom line. It was all about making money, not making beer. So, I lost interest in the big companies."

Kerwin became Big Rock's second brewer, initially working under Bernd Pieper, who'd been recruited from Zurich's Löwenbräu brewery in the lead-up to Big Rock's opening. "He was the only technical person they had and he couldn't do everything," said Kerwin.

By that time, the company had expanded considerably, following a series of events that rapidly elevated the Big Rock brand. A strike of unionized workers across the three big breweries soon after the Calgary upstart opened meant its beers were consumed more out of necessity than choice. The 1988 Winter Olympics in Calgary again put Big Rock in the spotlight, when the world had the opportunity to sample brews at the beer tent the company set up downtown, on Stephen Avenue. Thanks to a salesperson named Alistair Smart, said Shelagh, Big Rock also managed to infiltrate the event's VIP tents. After that, the beer writer and tastemaker Michael Jackson declared McNally's Extra, a rich Irish red long since retired from the Big Rock lineup, one of the best beers he'd tasted.

"McNally's Extra was a phenomenal beer," said Shelagh. "If you're ever lucky enough to taste it again, it's brilliant. And it appealed to all demographics."

With the increase in esteem came increased demand, and Big Rock was ramping up to open a $16.5 million, purpose-built facility in 1996 that was reported to have the capacity to triple annual production to 450,000 hectolitres and realize economies of scale it previously could not. Kerwin was to be the project's brewmaster, with a focus on improving efficiencies and quality.

The old equipment at the original site simply wasn't good enough to ensure batch-to-batch consistency, he said. The move "was a good opportunity to build a brewery from a greenfield site and put in the technology that would help Big Rock succeed."

Despite being a scale-up for Big Rock, it represented a return to roots for Kerwin. After years of watching automation eliminate unionized workers at his previous positions, "it was fun for me because we got back to making beer very basically, the way it was when I started. When you went to a craft brewery, everything was done in a very labour-intensive way. It basically got back to the fundamentals of brewing, which was kind of refreshing."

Unfortunately, that feeling would not last. Big Rock would remain independent, but it was showing signs of becoming something like what Kerwin had previously endeavoured to escape. The company had begun to eye market share dominated by the macro-brewers as indicated by new lagers meant to entice a typically non-craft audience. "I was becoming tired of the corporate culture that Big Rock was becoming," he said. "While Big Rock had been a great place to make beer, I was at the point where I wanted to do something different." While history was repeating for Kerwin, it was also repeating for independent brewing. A company was growing in a way that forced it to question whether it was time to move beyond its crafty origins. Alberta had seen that a century earlier, when another Calgary businessperson named A.E. Cross decided that he could meet the local demand for high-quality local beer. Like McNally, he did. Unlike McNally, he ended up playing a significant role in restricting the growth of craft beer in Alberta at the same time that it was booming elsewhere.

THE AMBITIOUS, TRUE PIONEERS OF ALBERTA BEER

Very few people know more about the early days of Alberta beer, or at least about A.E. Cross and his iconic brand Calgary Beer, than Spencer Wheaton. I met Wheaton through a Kijiji ad in fall 2018. I'm a casual collector of vintage beer glasses and one day I found what I thought was an interesting piece in an Edmonton second-hand store—a small glass bearing the Calgary Beer brand, a bison framed by a horseshoe. It was labelled ginger ale, which struck me as odd. Googling led me to Wheaton, who is no casual collector, to say the least. He keeps that ad posted on Kijiji in cities across the Prairies and Ontario. It's a standing offer to consider purchasing anything related to the Calgary Brewing and Malting Company. I sent him a note with a photo and he got back to me within a day.

It turned out that what I'd found was somewhat old, likely dating to the 1960s, but it wasn't rare. Not to Wheaton anyway. He had dozens of similar glasses. He also had a lot of other stuff related to the company founded by Cross, a local rancher and businessman (sound familiar?), just before the close of the nineteenth century. From the pictures he sent as we continued to exchange emails about the history of the brewery, which operated until 1994 as one of Calgary's longest-surviving businesses, I realized it was all much more interesting than an old pop glass. If I happened to be coming to Calgary sometime, Wheaton added, I could drop by his place for a look.

The following spring, I did. Wheaton lives in a modern, grey-sided townhouse in the city's south. Having just put his little boys down for a nap, he swung open the door before I had a chance to knock, and greeted me with a broad smile and a crushing handshake. He was a young-looking forty-year-old, tall and surprisingly thin, I thought, for being an executive chef at a local Italian restaurant. I took off my shoes and he led me to the door of the basement. "This is where the collection stops," he said, indicating the threshold of the door jamb as we began to descend. "Otherwise my wife is going to lock me out of the house."

After I'd explained a little more about wanting to learn about Calgary Beer, and its possible relation to the craft beer of today, Wheaton launched into a tour of the basement. One could compare the space to a museum, the only thing missing being the velvet ropes. Documents, pictures, and prints had been lovingly framed, and artifacts were protected in glass cases. A cupboard above the bar at one end of the room was full of old bottles that were in turn full of old Calgary Beer.

The collection, Wheaton explained, had started almost accidentally. A few years ago, a buddy came across some old Calgary Beer posters at a garage sale and Wheaton took one off his hands, just for something to hang on the wall. He became enamoured of the bison-and-horseshoe logo, which he now adoringly describes as "one of the most iconic of all time." With that poster, his collector's tendencies, which he's had since he was a kid, took hold. They've since manifested as all but obsession, checked by a spouse who, understandably, does not want to see her home transformed into a shrine to a long-lost beer. Nevertheless, Wheaton scours the Internet daily for new additions.

Some of the most interesting pieces in the collection were among the first he showed me, in that they suggested a direct link between the earliest days of Cross's company and what craft beer is today. In a curio cabinet along one wall were five empty bottles arranged in a V-formation. At the head was an empty green hand-blown bottle recovered from a barn in Crossfield, a town thirty minutes north of Calgary (and not named after A.E. Cross). Someone found three of them, said Wheaton, and then posted them to the Facebook group for enthusiasts of Calgary Beer memorabilia. He dates the specimen to the late-1800s. He did not make any kind of motion to take it out of the display case.

"It's probably worth about $600," Wheaton said.

The bottles on either side dated slightly later, and the ancient empties could today represent a foundation of modern craft beer, with labels indicating a stout, porter, pale ale, and even a malty German-inspired bock.

Next to that case, however, was an item that showed how significantly Calgary Brewing and Malting ended up breaking with those somewhat crafty origins. Hanging on the wall was a framed original lithograph of the Calgary Beer brewery, circa 1910, by Canadian painter and illustrator A.H. Hider. It was realistic to the point of being more documentarian than pretty, featuring everything from what appears to be the accounting room to the malting floor to the boiler room. It's a marvellous depiction of the incredible scale to which the operation quickly grew. It was no craft brewery.

Cross saw that growth coming well before Hider depicted it. Even by 1900, the company was advancing so steadily that the prospect of prohibition, creeping westward from Manitoba where citizens had voted the previous year to enact the ban, did not seem to faze the founder.

"It should not affect us very much," Cross wrote in an unfinished, handwritten draft of an annual report housed in the Glenbow Archives, "as it will either close out or weaken [our] principal opposition [so] that the lost trade will be counterbalanced."

The company rode out the dry spell by shipping beer to Mexico and by ramping up production of its ginger ale and other soft drinks. In another display case, Wheaton had a small collection of pop bottles, crystal clear and empty, that he believes date to the time of the ban on liquor that was in force in Alberta from 1916 to 1924.

Indeed, the company became a brewing juggernaut, employing hundreds of people by the early 1900s in a city that was then home to just over ten thousand. It expanded through acquisition as well, developing into a near monopoly by absorbing local brewing competitors such as the Golden West (in 1910) and Silver Spray (in 1925). Predating the oil industry of Turner Valley, which began in earnest in the mid-1930s, the malting and brewing company represented one of the largest local industries of the time. That, Wheaton told me, was a big part of what helped shape the business's reputation for being more than just a money maker. It also became an upstanding member of the community, supporting local athletics and building attractions and amenities.

"The Brewery under the Cross family was very community-oriented," says the Glenbow Museum in its description of the collection containing materials related to Calgary Beer. "It actively supported local sports through the Calgary Buffalo Athletic Association, developed the Brewery grounds as gardens for the enjoyment of Calgarians, and in 1938 established a Fish Hatchery. Two personal projects of [A.E. Cross's son] J.B. Cross were the Aquarium, opened in 1960, and the Horsemen's Hall of Fame (a western heritage museum), opened in 1964." And, of course, A.E. Cross was among the Big Four, the local business-men American rodeo promoter Guy Weadick enlisted to financially support the creation of the Calgary Stampede, launched in 1912.

Calgary Brewing and Malting, and the Cross family, are revered among the province's early brewers for their impact on their communities, but they were not alone in the effort. Among Cross's closest rivals was Fritz Sick, a German immigrant who built his first brewery in Lethbridge in 1901 and released Old Style Pilsner in 1926, the beer that accounts for the bulk of his legacy. Locally, residents of Lethbridge looking to keep fit would know the Sick name for more than that workhorse lager. In 1943, after years of buying and starting breweries in Saskatchewan, British Columbia, and Edmonton (his son, Emil, would also go on to expand the business into the US), Fritz donated $100,000 to the City of Lethbridge for construction of the Fritz Sick Pool. It continues to operate as the oldest place in town to take a dip.

"As well, the Sick family/brewery built the Brewery Gardens," Lethbridge Historical Society president Belinda Crowson told me by email. Located near the brewery, on the site of what was once a dump for the operation's coal ash, the gardens became a popular tourist attraction, Crowson added, though they were established under Fritz's son, Emil.

"The Sick home still stands in Lethbridge, but I don't suspect many people know which one it is."

In Edmonton, recognition of the names of its first brewers is even less likely. As in Lethbridge, the business began with a German immigrant who recognized an opportunity. Robert Ochsner and his wife arrived in Edmonton in the 1890s. In 1894, they built their brewery high on the south bank of the North Saskatchewan River, very near the current site of the High Level Bridge, and directly overtop of a natural spring that was ideal for beer making. It's still there—or some form of it, at least. I visited it on a brisk day in late April 2019, soon after seeing Wheaton in Calgary. The buds on the river-valley poplars that completely hide the facility in the summer were swelling but slow to open. These days, the old brewery is owned by the City of Edmonton. Road maintenance crews use one side of it, while the other is home to the City's historical artifacts division. Curator Benita Hartwell met me at the door and asked, cheerfully, what I hoped to see.

I'd soon discover that there was, in fact, not a great deal to see—not regarding beer, in any case. But it did help confirm a pattern I'd seen in Wheaton's collection. The old building was not the house that Ochsner built. He sold his in 1907. The buyers tore it down and replaced it with brick.

But very little of the "new" building, as it exists today, said *beer*. Hartwell led me through a room built as an expansion of the facility in the 1970s. It was about the size of an elementary-school gymnasium, and stuffed with all manner of furniture, equipment, fixtures, and more, all of it dating back decades and which were used in exhibitions in museums such as the Royal Alberta and Fort Edmonton Park. Hartwell and her colleagues were in the process of recataloguing the items to make them easier to find. "Fifteen thousand pieces to go," she said, smiling.

The one connection between this building and that of the Ochsners, and proof that I stood on hallowed beer ground, was the spring. At certain times of the year, it still surges, forcing its way into the building, much of the backside of which sits below the surface of the aquafied hillside. A central stairwell bears the water stains. There was even a section of the floor that was kept perpetually clear of antiques, as the water can sometimes pool there. "It's not so good for a museum collection," said Hartwell.

But the building did contain some relics that pointed to the evolution of beer in Edmonton, of how it would flourish in some ways and not in others. Early in the tour, Hartwell stopped at a large, glass-fronted cabinet and pulled out a Bohemian Maid bottle, the label worn, torn and barely clinging to the glass. It was the beer that was made in that very facility, by the North West Brewing Company, which revived the operation after Prohibition. There was also a deck of cards bearing a portrait of the fair maid herself, pristine in the original cellophane.

About halfway through the tour, we stopped to admire an ancient and desiccated wooden cask. Beside it was an aluminum one marked *Strathcona Brewing Company,* a leftover of what was the city's first small-batch brewery of the modern era of beer making. Started in the late 1980s, it borrowed its name from the operation that originally bought out Ochsner.

But the highlight of our walk through the archives came near the end: a mint-condition Yellowhead Beer sign. It was a white glass bulb slightly smaller than a soccer ball, and probably around a hundred years old. Likely, it was the product of Edmonton's famed "beer castle." The five-storey brick building, located a few blocks west of the city's downtown, looks precisely as it sounds, complete with ornate mouldings, arches over windows, and a turreted central tower. It was the result of the successive scaling up of William Henry Sheppard, local hotelier turned brewer. He bought his original equipment from a competitor of the Ochsners, the Yellowhead Brewing Company. The eponymous beer was an homage, which would be perpetuated by the modern-day Yellowhead Brewery when it opened downtown in 2006.

Hartwell held the glass piece and pointed out the carefully hand-painted lettering. It was beautiful in a rigorous way, indicative of a level of craftsmanship that has all but been replaced by technology. She put it before a nearby mirror, as had obviously been intended by its maker. Suddenly, the back of the sign, identical to the front but painted in reverse, was readable.

"It's cool how craft breweries today make connections to the past," Hartwell said. "It's like a resurrection of sorts."

Our tour ended soon after, but a thought stayed with me that also applied to Wheaton's collection. *Resurrection* was an interesting word for her to use, accurate in a surprising way. It pointed to an unexpected conclusion—the link between craft brewing of today and that of yesteryear is tenuous. Certainly, those old breweries showed the same kind of initiative and drive to succeed in an environment that was, to say the least, inhospitable. Alberta's first brewers built their operations from literally nothing. Unlike a modern brewery, the Ochsners had no municipal water line to tap into, so they built on top of a bubbling spring. When trying to figure out a better way to get his beer out of the valley from the beer castle's precursor, a brewery built not far from the banks of the North Saskatchewan River (Alberta's oldest standing brewery, incidentally, and now a residence), William Sheppard invested in the construction of a funicular. Fritz Sick ran his operation almost single-handedly, doing his own malting, brewing, selling, and distributing for years after startup. And A.E. Cross, the granddaddy of Alberta brewing, set the model for all of them, and showed that considerable wealth and economic impact could come of beer.

But in doing so, even if inadvertently, Cross and those earliest small breweries were largely responsible for killing small-batch brewing in Alberta.

What had started as Ochsner's brewery before the end of the nineteenth century ultimately ended up as part of Molson Breweries in 1989. Years after the brewery was revived by North West, it was taken over by Calgary Brewing and Malting. In 1961, Calgary Brewing sold to Canadian Breweries, an Ontario-based conglomerate that would morph into Carling O'Keefe. They'd run the brewery for more than a decade longer, before ultimately merging with Molson. Sheppard's beer castle would go the same way, in this case sold to Fritz Sick. Molson was waiting for him as well, it would seem. Sick sold in 1958.

Which means that true Alberta beer, at least in terms of corporate ownership, ceased to exist shortly after the mid-twentieth century.

It's not entirely the fault of those early brewers. Those small companies grew as the province grew, they worked to keep up with demand, increased their value, and then exited. That's simply the nature of conventional business. Nevertheless, they moulded the market into an inflexible form that would take years to reshape. (As early as 1909, incidentally, the province's first brewers, Calgary Brewing and Malting among them, signed a contract to sell beer at agreed-upon prices, indicating the level of control they had even then over the market. Today, that practice is known as price fixing and is an offence in retail markets in Canada.) Calgary Brewing started with styles easily recognizable today, but which were almost completely out of fashion for decades in Alberta. Maybe tastes changed. Or maybe consolidation allowed brewers to get away with creating a profit-maximizing preference for lagers made with fewer ingredients, and often with adjuncts such as rice to replace more expensive malt. Either way, by 1961, the preference for lager was almost 60 percent in Alberta. Total beer sales in 1970 showed an even more pronounced shift, with lager accounting for 96.7 percent, ale 2.8 percent, and with porter and stout (though also ales) being so maligned as to have been separated out in surveys of the day, in which they barely held on at 0.5 percent.

What's more, the regulatory environment that evolved in concert with consolidation virtually eliminated the possibility that a small brewer might try to meet the needs of that non–lager drinking 3.3 percent and perhaps even grow that market so that other small brewers might get in on the action. The Alberta Liquor Control Board implemented a 500,000-litre annual minimum capacity requirement, limiting a potential beer-maker's options to mass production-sized brewery or nothing at all. Staff members at what's now Alberta Gaming, Liquor and Cannabis don't know when the rule was set, or precisely why, other than that it was possibly intended to limit the business to large practitioners who ostensibly would have had more reliable food safety practices. Conspiracy theorists might say it was result of pressure from ensconced industry members uninterested in sharing the market. In any case, if the minimum capacity requirement didn't deter an aspiring brewer, the cost of a licence might. In 1958, the board charged $5,000 for one. That's around $45,000 in today's dollars. The cost of a brewer's licence in 2019 was $700, if you take into account the $200 non-refundable application fee.

Back in the basement of his south Calgary townhouse, none of this seemed to matter to Wheaton. Calgary Beer isn't actually dead, he pointed out to me, even if it stopped being sold regularly in Alberta in the mid-1980s. Now brewed by Molson Coors, it made a brief reappearance in 1992 and again in time for the hundredth anniversary of the Calgary Stampede in 2012. Then it was gone again. These days, Wheaton asks friends travelling through Saskatchewan, where it can be found on liquor store shelves, to bring him a six pack. It's mostly for novelty's sake.

"It's just a generic lager," he said with an unconcerned shrug. To him, how the company ended up is almost inconsequential compared to what it once was. "They really were an integral part of Calgary," Wheaton told me. "When people look back [on the brewery], they don't just remember the beer. They remember what they did for the city."

In his mind, what it is now, or isn't, doesn't change that.

"THIS ISN'T WHAT I SIGNED UP FOR"

In 1997, Big Rock went public and began trading on the Toronto Stock Exchange. In terms of entrepreneurship in Alberta, it was a Cinderella story; after years of toil, the company was finally getting a well-earned invitation to the ball. It began taking bold steps that would see the company experiment with going head to head with the big producers. It would launch Big Rock Lime, for example, as a response to the trend for citrusy lagers in the late 2000s. Soon after, a lager called Gopher was brewed as an alternative for the rare Molson Canadian drinker who sought a local alternative. Eventually, the brewery came to its senses.

"We are a craft brewer and we needed to behave like a craft brewer and we are now," said former president and CEO Bob Sartor in 2014.

It did begin to behave accordingly; in time, it launched a barrel-aged series and a sour program that would garner regional and national awards. All the while, of course, it never stopped brewing its Traditional Ale.

Shelagh McNally remembers the transition from her father to Sartor in 2012 and the changes in strategy that came with it as though it were a personal loss. Nostalgia weighed heavy on her heart. After all, she'd started working at the brewery after graduating from the University of Alberta with a degree in political science. Though she started out doing whatever was asked of her—anything from cleaning out the bottle-labelling machine to delivering beer to answering phones—she'd end up working very closely with her father, so much so that it's unclear to her today whether some of the ideas behind the company's success over the years were hers or her dad's.

"I came into the brewery [after the change] and I was looking around and Ed wasn't there," she said. "This isn't what I signed up for. Dad and I had so much fun. And when he wasn't there, it just wasn't... Beer is supposed to be fun, in my mind. Beer is something that brings people together. If it's not fun, why the hell do it? It had become a business. And you have to be. You have shareholders. But, I don't know..."

Shelagh is no longer involved in the company. Today, it's focused on remaining accountable to its shareholders in a way that goes beyond that reexamination of identity under Sartor. Its 2019 first-quarter results included a loss of $1.7 million, and it pointed the finger at the previous NDP government that had endeavoured to position itself as a partner of a burgeoning industry. A tax break that that government put in place for new and small brewers did not apply to Big Rock, which was too big to qualify. At the same time, however, it was effectively penalized for its success—it's not yet big enough to warrant being treated like its multinational competitors, yet it was being asked to pay similarly high taxes on each unit of beer produced.

"The 104% increase in the net Alberta beer markup (provincial tax) imposed on Big Rock by the previous Government of Alberta in late 2018 (being a 160% increase since 2016) has forced the Corporation's senior management to take immediate cost-cutting measures," read a press release on the company's website on May 29. It went on to say that Big Rock looked forward to working with the current government on a solution. Layoffs followed.

There's no way of knowing if Ed McNally could have done anything to prevent that, were he in charge of the company at the time. All the same, Shelagh felt that he was an excellent businessman. That acumen was highlighted as a reason he was named to the Order of Canada in 2005. So was his philanthropy, in the form of support of the arts and post-secondary education. Shelagh also remembers and admires her father as "a great humanist." Growing up with relatively little during the Depression had made him "classless," she said. "He listened, and he wasn't an arrogant man." He never pretended to know something he did not, and nor did he pretend to be something he was not.

"He'd wear his sweater inside out and backwards and have dirty knees when he went into work, because he'd be weeding his beloved asparagus patch," said Shelagh. He played golf, but not solely for the love of the game. "He didn't compete. He just wanted to have fun. He just wanted to enjoy the people. So, beer was so perfect [for him].

"He was a character," Shelagh summed up. How much she missed her father was obvious in her voice, faltering slightly now and again as she reminisced. "He was a great guy."

For Kerwin, the changes at Big Rock were a sign that it was time to move on. Five years before McNally stepped down as CEO in 2012, Kerwin would leave to explore new paths in craft beer. One of those would ultimately lead him to co-found Calgary's Village Brewery.

But if history had repeated, no one could deny that it had also been made. Because of one man's desire for what he considered a good local brew, craft beer as we know it in Alberta today was born.

IN ED'S FOOTSTEPS

The Ed McNally of Edmonton might be Neil Herbst, but one of Herbst's great regrets is that he never got the chance to sit down with McNally and have a beer. "I wish I had," he told me over a pint of double IPA one chilly fall day at Alley Kat Brewing Company, Edmonton's oldest operating craft brewery. "It was one of those things that we never got around to. Which is too bad. We didn't have a whole lot to do with them, oddly," he added of Big Rock.

But then, who had the time?

In 1995, as Big Rock was moving to a new facility and soon to go public, Alley Kat was just opening in a business park in south Edmonton, a few blocks from the massive Labatt brewery. It was just Herbst and his business partner at the time, Richard Cholon, trying to fill what they saw as a gap. Craft was not new to Edmonton back then, but it was unfamiliar to most drinkers and, because of that, shaky as a business model. It would remain that way for nearly two decades to come, with Alley Kat outliving other local craft breweries who'd come and go.

"We thought there was room in the market," said Herbst. He'd taken a buyout from a job with the provincial government's apprenticeship training division and hoped to parlay his experience as an award-winning homebrewer into something bigger. Flanagan & Sons, which also operated on Edmonton's south side, was then the only craft brewery in town as Herbst and his partner were formulating their plan. "We had an assumption that we could do better," he said.

Early signs indicated that taking a crack at small-batch brewing might not be that hard after all. Meeting the minimum brewing capacity, for one thing, turned out to be a matter of mathematics involving the twenty-hectolitre system with which Alley Kat started out. Herbst recalled hearing that brewers in Calgary were often required by licensers at the Alberta Liquor

Control Board to actually brew that minimum or better. In his dealings with the authorities, he merely needed to prove that it could be done if they did so many brews each week.

"They were pretty easy going on it," he said. "We fit the provincial requirements." But as far as actually making and selling 500,000 litres of beer in twelve months, "it took us years to get there." That was not exactly what they expected. "In our business plan, we were shovelling the money in after about six months," said Herbst with a laugh. "In part, we didn't really know the market we were getting into."

He and Cholon had looked at the thriving micro-brewery scene along the American West Coast and British Columbia when they got down to committing their ambitions to paper.

"We assumed the uptake would be the same here. What we didn't realize is that Alberta didn't get as much of its influence from the West Coast as it did Central Canada."

Central Canada, of course, was the birthplace of the likes of Molson and Labatt, both of which were kept busy in Edmonton with the relentless task of meeting the demand for easy-drinking lagers. Herbst recalled Alley Kat's first hop-forward offering, its flagship Full Moon Pale Ale, was originally brewed on contract for a now-defunct Edmonton brewpub called Taps. He and Cholon liked it so much they decided to brew it for themselves as well. But their appreciation for it was not widely shared. "When we came out with Full Moon, it was 35 IBUs [International Bittering Units] and people were like, 'Oh, that's . . .'" Instead of finishing the sentence, Herbst made a face as if he'd just smelled something bad. To compare, a bottle of Labatt's Blue doesn't crack 10 IBUs.

Even a tame wheat beer that was among Alley Kat's early offerings was at times met with consternation or confusion. "We thought people would understand what it was but often we'd get it back," said Herbst. "People would say it's cloudy, there's something wrong with it. It was a bit of a slog." As a result, the brewery took on the role of both educator and heretic. "When we started off, we assumed that if you brewed it, people would buy it. You realize that, no, you have to go out and sell it. You have to educate people and you have to make sure your beer is top notch."

For about three years after it started, Alley Kat made no money, Herbst said. Around that time, his wife, Lavonne, bought out Cholon. She put her MBA to work on business-building while Herbst kept the tanks full. "We figured out how to exist in that sort of market, at relatively low volumes. It's not like we were packing mattresses full of cash. It was not super profitable. But we could make a living at it."

But even before the craft beer boom, "you could see the market moving a bit," he added. That and "doggedness" kept him in it. Alley Kat has grown steadily ever since. That's measured in single-digit percentages these days, but that's fine by Herbst. Alley Kat will likely never be another Big Rock in terms of size, but that was never the plan anyway. Too much growth, he said, would have meant too much strain on employees, and on him and Lavonne. In that sense, it has remained manageable. Now in his early sixties, Herbst—known by many in the industry as the "Hop-father"—says he has no immediate plans for retirement. He doesn't do the brewing at Alley Kat anymore, but he still shapes the product and the direction of the company by working on recipes on his brewing system at home.

"I've gone full circle," he said.

Herbst may not have had the pleasure of a beer with Ed McNally, but he considers McNally to have been instrumental in establishing the craft beer industry we know today. And even if Herbst did his fair share of educating consumers, and likely laying the groundwork upon which new brewers can more easily build today, in the post-2013 era of craft beer, he never speaks of himself or of Alley Kat as having had a significant impact on the industry today. (Although in 2019 his self-deprecation was upended at the Alberta Beer Awards, where Alley Kat was named brewery of the year.) Herbst did, however, have an auspicious meeting with another of the founders of the Alberta beer industry. As he was preparing to open Alley Kat in the mid-'90s, Bob Blair, who married into and worked for the Sick family, stopped in to see how the brewery was coming together.

"He brought us a bouquet of flowers and an old, original Sick's beer glass—a ceramic beer glass—with a note," said Herbst. "I never asked him what exactly he meant, but the note said, 'May you be half as successful as the Sick family.'

"I didn't know whether it was a curse!" said Herbst with a smile. "He was a pretty funny guy."

WHILE HERBST SLOGGED away in Edmonton pre-2013, efforts to convert the masses also carried on in Calgary, with upstarts that included Wild Rose Brewery, whose success would lead them to be acquired in spring 2019 by Sleeman Breweries. But proselytizing in the name of craft beer wasn't limited to the major urban centres. And it certainly wasn't any easier.

Kevin Wood set up a brewery in Red Deer in the late 2000s. Like Herbst, he felt that an opportunity existed in a province with only about half a dozen breweries at the time. It did, he found, but only to a point. "We opened a fairly large brewery," he said, with about $1.5 million cobbled together from friends, family, and private investors. "It was very difficult. It took us about two years to get the equipment together. There were no manufacturers around."

At first, the plan was to start with craft beer. Despite brewing 860,000 hectolitres "right out of the gate" in 2009, he wouldn't quite get around to establishing himself as a craft brewer until about five years later, when he wouldn't need to brew that much anyway. Instead, he built his business not by trying to educate the masses as Herbst had, but simply by giving them what they wanted: a plain lager, and one that already had a reputation.

That lager was Drummond, which had been previously brewed in Red Deer until the brand was retired in 1995. Wood revived the lapsed trademark and even brought back the beer's original brewer to get the recipe right. "I remember Drummond as a kid," said Wood. "My dad drank it. Everyone drank it."

During our telephone interview, Wood seemed to be agnostic about the brand back then. To him, it just made good business sense. "Some people drank Drummond, some people thought it was the worst beer in the world. At the end of the day, I would rather have had something someone had heard of than nothing. Because it was very hard to get market penetration back then."

Biding his time, Wood left the Drummond brand untouched. He knew what his customers expected of it and he wasn't about to lose them. "You can't do a Drummond IPA or a Drummond white ale," he said. "When the consumer buys [that], they're going to be like, 'What the fuck is this?'" Then, in 2014, when the market was poised for an upswing, he made an addition to the house that Drummond built, and launched the Something Brewing brand with four core beers: a wit, an IPA, a nut brown ale, and a schwarzbier (a German dark lager). The relationship between the two brands is an open secret.

"I try to keep the two separate but everyone knows Something Brewing is brewed at Drummond," said Wood, unbothered.

For now, that works. One day, he'd like to have a small brewery dedicated to producing those four core offerings and a series of one-off specialty and seasonal beers. In the meantime, he's still building the business, and illustrating the reality of the beer business for the next generation. Wood remembers a time when he had to go looking for a $100,000 bailout to keep the operation going that same afternoon—and pulling it off. He still hustles, it seems, for every dollar. Wood said that he puts almost every dollar made back into the company. Unlike Herbst, he's more overt about aspirations that seem at least in part inspired by Ed McNally's ambition, if not by the brewer's iconoclastic approach to beer making.

"Right now, I have half a percent market share, so one in every two hundred beers in this province is mine," said Wood, referring not only Drummond but to Beer Beer, a brand he also revived.

His hope is that the strength of those humble beers, with Something Brewing in the wings, will see that half point grow into one within three years. He's also looking beyond the province. Currently, other provinces do not enjoy the same open retail market as Alberta. Elsewhere, governments, instead of private agents such as importers, control which beers get shelf space at liquor stores. When the day comes that those barriers are finally removed, Wood believes that his experiences with near-mass-market beer will prove invaluable.

"Once the other provinces start privatizing, I'm going to jump in there right away," he said. "And I think I'm going to be at an advantage because I know how to operate in a free-market economy."

WE'RE IN AN era now, of course, in which any other kind of market is fast becoming veteran's lore, histories akin to tales of having had to walk to school uphill both ways, and always in the dead of winter. But the pre-2013 era was marked by absurdities, a mishmash of impractical laws involving the kind of government oversight that was simply a holdover of post-Prohibition morality around alcohol. While it would have been justified as part of dealing with a controlled substance, it was invariably an impediment to entrepreneurship. For some businesses and industry members, that led to trouble they'd never have expected. Take Don McDonald, for example.

Government regulations meant that, twenty years ago, he ended up learning how to brew by correspondence.

I met McDonald in late 2018 at the Brewsters Brewing Company location in Oliver Square, a shopping complex at the edge of one of Edmonton's hippest neighbourhoods. We each had a pint of the smooth but tangy Mango Milkshake IPA that was among the brewpub's latest releases at the time. By then, McDonald was an industry veteran with one of the oldest craft companies. He'd been with Brewsters for about twenty-three years. Brewsters itself had got its start seven years earlier, in 1989, when founder Michael Lanigan opened the first one with his brothers, Marty and Laurie, in a Regina strip mall. McDonald started out in Calgary, as a day manager. Now, he was director of sales. It was a stepwise evolution.

"I've done virtually every job there is to do in the organization," he said. The reason that included brewing was due to a market that in no way approached Wood's notion of "free." As a brewpub, Brewsters was exempt from the 500,000-litre minimum, but it could not transfer beer it made at one location to another. Therefore, each pub had to have its own pricey, space-eating brewing equipment and a qualified brewer to operate it. When the brewer suddenly left his post at the Lethbridge location in 1998, McDonald, there as a manager, was called upon to step in. Though the laws relaxed that year to allow beer to finally be transferred between locations with the company, the equipment was already there and the decision was made to brew local beer for local drinkers. He called the head brewer, Robert Walsh, in Edmonton.

"I'm like, 'Rob, I'm running out of beer,'" recalled McDonald. "And he's like, 'Well, you have to brew some.' He literally taught me over the phone. I brewed for about six weeks."

The laws that kept brewpub beer from being moved among a company's own locations extended to retail until 2013. The beer could not be sold in liquor stores, further hamstringing businesses that, by law, had been required to duplicate their means of production.

Looking back, McDonald can actually see the positive in it all. The situation forced the company to focus on brewing just for its own "ecosystem," he said, dialling in the appropriate volumes and varieties, with room for some experimenting on the side. Walsh was brewing traditional European "farmhouse" styles such as saison and bière de garde as far back as 1999, said McDonald. Brewsters's annual holiday release, Blue Monk—a strong, rich ale categorized as what's known as a barley wine—dates back nearly to the brewery's founding. (The milkshake IPA was still a few years off.)

From a salesperson's perspective, however, not having the variable of retail external to the company thrown into the mix simplified the process. "I didn't have to worry about 'Are the liquor stores going to order this week?' because I know they're not. In some ways, I think we were protected from the peaks and valleys of the industry."

Today, the company has centralized production of its core beers at a facility in Calgary that expanded incrementally as the market dictated. It focused production of one-offs and seasonals at the smaller system in Oliver Square. And McDonald was eyeing growth. When envisioning the future of the brewpub, he looked in some ways to Big Rock. He did not mean size, though; it was brand recognition he was after. At the time of writing, Brewsters beer could be found in eight hundred liquor stores in the province (out of roughly fifteen hundred), McDonald said, as well as some in Manitoba. He'd like to see Brewsters beer in every shop in the province, of course (and, one day, Brewsters brewpubs across Canada, but one thing at a time). "If you're opening up a liquor store, you're not going to open up without Big Rock. You're going to have Trad and Grasshopper on the shelf. We want to be the beer that is recognized like that.

"We still haven't finished Alberta yet," he added. "We've got a lot more that we can still do here."

NOT EVERY STORY is as optimistic as that of Brewsters. There are those for whom the legislation and other market factors were death knells. Among the ghosts of Alberta's craft beer pioneers is the Strathcona

Brewing Company, a small company in which Kathy Fisher had invested.

"We never stood a chance," she told me.

I met Fisher one day at her south Edmonton home, not far from the site of the old Bohemian Maid brewery on the edge of the river. Hers was a compact, well-loved bungalow that looked like an original in a neighbourhood giving way to boxy infill. The front porch slumped lazily toward the sidewalk, and a faded string of what looked like Tibetan prayer flags fluttered in the breeze. Inside was an organized chaos of stacks of books, shelves of CDs, and pictures that panelled the walls. Fisher is a tall woman with long grey hair that she often pulls into a ponytail, eyes made permanently squinty from an unrelenting smile, and a manner of speaking in which one profanity-punctuated sentence crashes into the next. She led me to the kitchen where she showed me things she'd described at a friend's potluck a few months earlier. They were relics from her brief relationship with Alberta craft beer.

"The upside of being a hoarder," said Fisher. "No one will have any of these."

On the Formica counter sat a pint glass bearing the logo of the Strathcona Brewing Company, freshly wiped of a layer of dust. Beside it was a six pack of pale ale from the brewery, bottles and box unopened for roughly a quarter-century. Fisher was never a brewer. She's a lawyer (not to mention a poet and filmmaker).

But having grown up in Montreal and the Eastern Townships, beer was always a part of life. Her grandmother was prescribed Guinness, rich in B vitamins and iron, after successfully giving birth. With a bit of a giggle, Fisher remembered drinking beer après-ski in her mid-teens. As it turns out, these would be happier memories than some of those she'd make during her attempt as a trailblazer in Alberta craft beer.

The trouble wasn't the beer itself. Fisher remembers Strathcona Pale Ale, originally brewed by co-founder Tom Daly, as being of top quality. Alley Kat's Neil Herbst mentions Strathcona—Edmonton's first microbrewery, started in 1986, nearly a decade before his own—in the same breath as Big Rock when talking about early influencers of Alberta craft beer. But, like most nascent breweries in the oppressive regulatory environment of the day, they were undercapitalized. That's where Fisher came in.

At the time, her partner was a man who was involved in managing the brewery (he "felt strongly," she later told me, about not discussing this time in his life with me). The business needed a canning line, and she had $50,000 as an advance on an inheritance from an aunt. In addition to that investment, she'd occasionally do sales on weekends.

"It was easier than selling legal services," she recalled. "The pleasure factor is there with beer."

But, in the end, the margins proved too thin, and the business atmosphere was inhospitable. Strathcona struggled to get a decent supply of malt that wasn't mouldy and unusable. As well, just like Herbst would find, the market was immature. Strathcona was ahead of its time at a time when that was a bad way to describe a brewery.

"I found it sad," said Fisher, who ended up getting about half her money back by selling off shares in the business and writing the investment off as a business loss. "Tom had a great vision. It's like you write the best book, [if] your audience isn't ready, you're fucked. We were a little early on the scene."

After the company went into receivership in 1993, its equipment was purchased by Flanagan & Sons at fire-sale prices. ("We hoped to just break even," said Sean Flanagan, who went in on the venture with his wife, Joyce Lester. But they, too, would not survive to see the day when craft would catch on in the capital city.)

Other than knowing that the founders' dreams went unrealized, Fisher has no regrets about having taken her shot on Alberta craft beer. "The business of beer is one tough racket. It was far more complicated than I had ever imagined," she said, looking back on the episode philosophically. "It's not what a thing cost you. It's what you learned."

Similarly, Fisher doesn't envy any of those who made it through, nor those who made it big. She doesn't spend much time wondering whether Strathcona could have made it today, in our current deregulation heyday, twenty years too late for Edmonton's first true craft brewery, but just in time for a new generation of brewers and a market that's finally ready to embrace them. Any of those new brewers, of course, will still say that beer is a tough racket, but at least there are now fewer odds to beat.

As our visit drew to a close, Fisher returned the glass to its cupboard and lugged the six pack back down to her basement, where it could conceivably never again see the light of day. I found it interesting that, with all the art, trinkets, and memorabilia, she did not keep at least the glass among them. Before I left, our conversation turned briefly to that other lawyer who endeavoured to introduce his city to craft beer: Ed McNally. Why had Big Rock worked when others failed?

"He had enough capital to see it through that initial period," she said, "and consistent management from the beginning." There was no tone of resentment in her voice, though perhaps some wistfulness. There was also a great deal of respect in it. Fisher appreciated that, even if Daly couldn't, even if *she* couldn't, someone made good on their plan to turn Alberta on to craft beer, or at least plant a flag until things changed enough for others to join the fight. At least someone beat the odds.

"I look at Big Rock," said Fisher, "and say, 'Bravo, Ed.'"

2

JUST LIKE STARTING OVER

*Setting the stage for
Alberta craft beer*

FOR GRAHAM SHERMAN, December 5, 2013, came like the sound of a starting pistol for the race, an ultra-marathon of a beer run, if you like, for which he'd be training for years. The day had finally come: Alberta's Progressive Conservative government announced through the AGLC that it was eliminating the minimum brewing requirement that had been in effect for decades. Before then, beer makers could qualify for a licence from what was then the Alberta Liquor and Gaming Commission only if they demonstrated capacity to produce 500,000 litres of beer annually.

"That is in the top 5 percent of breweries in North America," Sherman told me. To him and business partner Jeff Orr, whom he met in Afghanistan when they were IT consultants for the Canadian military, it seemed ridiculously restrictive. "We got looking at it from the standpoint [of], how on earth could this be legislation in this province when we grow the best barley? It doesn't make any sense."

So, with ambitions to use that barley to build a Calgary-based business they called Tool Shed Brewing, Sherman and Orr railed against that legislation, taking their arguments to government and their gripes to the media in an effort to change rules that were stunting the development of an industry. Few would attribute so momentous a change to a couple of guys making a racket about how pointless the legislation seemed. Nevertheless, Sherman, a born and boisterous storyteller who's no stranger to the international speaking circuit, crafted a compelling case, one he began making not long after he and Orr made their first batch of homebrew in Sherman's eponymous backyard tool shed in February 2012.

At first, they had no intention of starting a brewery. All they wanted, said Sherman, was to make beers they liked but that were hard to get in Alberta—hoppy and complicated and flavourful stuff inspired by the likes of great American craft brewers such as Dogfish Head, Oskar Blues, Sierra Nevada, and Lagunitas. But making and sharing that beer led to an epiphany.

"Dude, this is it," Sherman told Orr one day. "You know what we do right now? We make people happy."

He knew it was the thing he wanted to do for the rest of his life. The question was how. The answer, at least in the short term, would be ludicrous.

Opening a brewery of the size the Alberta government required at that time was not an option for those without easy access to an awful lot of cash. Sherman recalled looking at Village Brewery, which sold its first keg just before the end of 2011, and thinking of the operation as being in a class of its own. It was. Each of the six founding partners of the well-respected Calgary brewery cut their teeth in Alberta craft beer after being long-time employees of Big Rock.

"At that time in Alberta, business in general was booming," said Larry Kerwin, Village's first brewmaster. "There was a lot of money around, so we very quickly raised enough to buy equipment [and] start up."

Fifty "beer barons" bought in as part of a limited partnership. "It was exciting and romantic in those days," said Kerwin. "Most guys had thought that owning part of a brewery was one of their lifetime dreams. They just wanted to be associated with a brewery."

No doubt, those guys also had confidence in the Village team, which included not just Big Rock's former brewer, but its chief financial officer. Tool Shed, on the other hand, was just a couple of guys who, as smart and resourceful and driven as they were, had little more than a serious passion for homebrewing.

"Those are the types of people who could go to the bank and to the community and finance a project like that," said Sherman of the founders of Village. "But for the rest of the world . . . it was virtually impossible."

With that, Tool Shed embarked on what now seems an exercise in absurdity. If they couldn't make beer in their home province, they would bring it in after making it themselves someplace else, in this case Dead Frog Brewery, then in Aldergrove, BC, with whom Sherman established a relationship to brew on their equipment.

"Essentially, we're shipping Alberta barley across the border, so we can brew in BC, and then drive home and import the beer back into the province. That's what it took," said Sherman.

During Tool Shed's time abroad, he made the round trip once and sometimes even twice a week, putting some 500,000 kilometres on his Tacoma pickup. By 2013, they had their first beers on the shelves of Alberta liquor stores. All the while, they were telling the Tool Shed story to anyone who'd listen. Sherman and Orr made regular appearances on the CBC, Global News, in the *Calgary Herald*, even popping up in the *Timmins Daily Press*. The stories tended to focus on the passion project and a hobby that had been taken way too far. And why not? It was a fun story. When they got home from Kabul, the pair had taken a deep dive into coffee, setting up a roasting system that could be controlled via Bluetooth. Barbecuing took a similar turn—the pit, connected to various sensors and monitors, would tweet Sherman when the ribs were done (there's a certain showiness there that captures

Sherman's personality perfectly, the idea that someone might be able to follow the Twitter account of his barbecue). Now they were nerding out on beer, applying similar high-tech solutions to dial in processes with an engineer's exactitude, and ready to scale up. All they needed was a bigger tool shed.

The story that Sherman really wanted known at the time, however, was that the Alberta government did not appear to want him and Orr to be anything but homebrewers. "The current system makes you want to leave the province to start a business," Sherman thought at the time. It required raising a couple of million dollars just to open the doors and Herculean efforts thereafter to move a great deal of product. For him and Orr, of course, that was bad. For the province, however, Sherman felt that it was an enormous missed opportunity.

"This isn't just for Jeff and I to get our buddies drunk at Tool Shed," he recalled emphasizing at the time. "I think there's a burden on all of our shoulders to stand up for something that would be great for this province, great for the economy, great for creating jobs, great for the farmers."

Look at wine, he told me, and the tourism industries that revolve around places such as California, France, Italy, and Germany.

"The common denominator is, you go to where the best grapes grow. So, for [Alberta], we've got the best barley but you [couldn't] brew here! It made no sense.

There was really no love for the amazing barley that's going into the world's best beer. Going to AGLC about it, we could never get an understanding of why that [minimum brewing] law was in place. It was mostly just kind of, 'Well that's the way it's always been.'"

(AGLC representatives, who did not wish to be quoted, later told me something similar. The rule, my source believed, was likely put in place based on the assumption that only a large company could be trusted to ensure that proper food and beverage safety standards were upheld. She did not know when it was implemented, only that the regulation had been around for decades.)

A few of those miles on Sherman's Tacoma also went to trips to Edmonton for meetings with the AGLC. They went in with data. Sherman leaned hard on a 2013 Conference Board of Canada report that detailed the economic impact of brewing.

Even then, the "beer economy," as the Board described it, was a major contributor to the national GDP. In fact, it ranked fourth among Canada's major industries, slightly behind the postal service and comfortably ahead of radio and television broadcasting. Out of every hundred jobs in the country, one was associated with beer. In Alberta alone, which the report did not consider a major producer at the time, nearly thirteen thousand jobs were linked to brewing, whether directly or through sales, food and hospitality, arts, entertainment, and recreation.

"The beer economy generated $5.8 billion in tax revenues annually on average over 2009 to 2011," the report went on to mention. What's more, "For every $1 spent on beer consumption in Canada, $1.12 is generated in real GDP." Throw a few more breweries into the mix, Sherman figured, and imagine the return on the investment. Soon enough, the AGLC did begin to consider the impact of an expanded industry.

"The AGLC did hear from quite a few stakeholders that this is actually making it hard to start a business," a spokesperson was reported saying in 2013, "and eliminating minimum production capacity requirements would allow Albertans to start up a manufacturing business or would allow current manufacturers to pursue other business opportunities."

Alain Maisonneuve, vice-president of liquor services at the time and now CEO, effectively announced a new role for the organization. Once called the Alberta Liquor Control Board, its emphasis was on *control*. Decades earlier, the organization even dictated the size and format of business cards of industry reps. "It is suggested that a well-designed, conservative business card is the most acceptable," read the regulations. In another example, a terse missive dated 1966 was sent to chastise the manager of the Calgary Brewing and Malting Company for not sending a mockup of a label change to the regulator for review prior to release. As of 2013, a new era of partnership was dawning.

"The craft industry is really growing in North America," said Maisonneuve, "and we're really excited about creating that opportunity in Alberta for Alberta businesses."

The day arrived. December 5, 2013. The AGLC announced that it had removed the 500,000-litre minimum. In its place was a new regulation, in that there were no longer regulations about how much beer a manufacturer should be able to make, be it a little or a lot. Instead, brewers would just have to meet federal standards around things like labelling and alcohol content, make beer of such quality that it would not make drinkers ill, and, perhaps most importantly, pay all taxes associated with alcohol production. It changed everything, instantly. For the media, Sherman was the go-to source for a response to the changes. "This is going to open the door for so many craft brewers—it's incredible," Sherman told the *Calgary Herald*. Finally, the Tacoma was about to get a rest. And, he added, "this means we can come home."

GOODBYE TO A "DUMB RULE"

Looking back on it now, it's clear that this regulation change was the fulcrum point for the craft beer industry in Alberta, but at the time, ironically, aspiring brewers such as Tool Shed were somewhat alone in their hopes for changes in legislation. The Alberta Small

Brewers Association had been created just prior to the shift Sherman and Orr pursued, but the minimum production change was not high among their priorities in the early platform they'd put before the government.

"The minimum requirement was on the list," recalled Greg Zeschuk, the organization's executive director at the time, "but it was way down the list, like number seven."

The association was more intently focused on the concerns of its current members, not of breweries that were mere glimmers in the eyes of brewers on the outside looking in. Instead, priorities included revising rules to make taprooms, virtually non-existent then, possible as the personable, brewery-focused places for a pint we know today.

The association was also focused on making changes to the mark-up schedule, the province's alcohol taxation structure that seemed to prohibit growth of existing small breweries. Back then, its defining feature was a "cliff," said Zeschuk. In comparison, income tax is graduated: A lower percentage is applied to a lower bracket of one's income, a higher percentage to the next, and so on. With beer, the rate of taxation was not stepwise. Once a brewery's total production reached the next defined level in the mark-up regime, its entire output was retroactively taxed at that rate. This created a "disincentive to grow," said Zeschuk, who would go on to build two breweries of his own in Edmonton, a brewpub called Biera, and another

nearby facility called The Monolith, which is dedicated to barrel-fermented beers.

At the time, the association was in a position to do
only so much. For one thing, it was small, as it preceded the brewery boom. For another, while it did not
have a strained relationship with government, the relationship was somewhat one-sided.

"We'd be like, 'Hey, here's a bunch of stuff [we'd like
to change]' and they'd be like, 'OK, got it.' Doors closed
and then they'd have a conversation," said Zeschuk.

Barb Feit recalls those conversations to have been
held in depth. Feit was chief financial officer at Big Rock
when she helped to found the Alberta Small Brewers
Association (she also served as the brewery's interim
CEO between January and September 2017, after Ed
McNally's successor Bob Sartor left). In 2012, she said,
the government began looking at the brewing industry
in earnest, and the AGLC opened a "beer file" she feels
hasn't really been closed since. The regulator began
meeting with brewers and discussing issues that were
seen as impediments to growth, compared these to best
practices in other regions, and eventually came up with
thirty-nine recommendations for policy amendments
around liquor manufacturing in Alberta. The one that
stood out the most, of course, was the elimination
of the rule that effectively excluded small producers.

"We knew that was the biggest hurdle for new breweries coming in and I think it's still the biggest reason
for the increase [in the number of breweries]," said Feit.

Indeed, the association had sympathized with those who were dissatisfied with the minimum production requirement. "Other new startups were like, 'This is a dumb rule. You're just discriminating against small business,'" said Zeschuk, echoing Sherman's sentiments. Like the AGLC, Zeschuk didn't know the origin of the rule, but wondered if it involved the influence of larger brewers.

"I think it was the big guys [who] somehow shoehorned this in under the guise of, 'It's food safety.' One thing that's interesting about beer is that it's exceptionally hard to make beer that's unsafe to consume. Beer is designed ... to be inherently safe due to its pH, alcohol, sugar-content. This is all anti-microbial stuff."

Whether or not it was founded on unwarranted health concerns, the regulation created "an artificial barrier to entry, a capital barrier," said Zeschuk. After having set up his own operations, he placed that barrier at a minimum of half a million dollars. Even that, however, "is not an easy thing to get," said Zeschuk.

One day before it happened, the association was notified that the industry was getting "number seven" on its wish list—the request to remove the minimum production requirement.

"Boom, out of the blue," said Zeschuk, "[the law] was gone."

Alley Kat founder Neil Herbst wonders if, to some degree, the barrier was more psychological in nature. Even after the minimum was unceremoniously struck from the books, opening a brewery would still cost most new brewers hundreds of thousands of dollars. It was expensive, he told me, looking back, but "it was not onerous." The real obstacle, Herbst felt, was the requirement to modify one's mindset and get past the thought that you'd need to not only make roughly a million pints of beer but convince people to buy them all—and do it while competing with some of the biggest companies in the world. That consideration, it seemed, would be a matter of faith, not finances.

Whether it allowed for more faith or easier financing, or both, the announcement in December of 2013 triggered an eruption. "Everything went wild," said Zeschuk. "There was a six-month lag phase and then the industry exploded. What I'm amazed at," he added, "is how did so many people have the same idea at the same time?"

BREWS NEWS YOU CAN USE

Zeschuk's question is valid. Was it the removal of one regulation that unleashed a tsunami of small-batch suds? Or was there more to it? Craft beer was definitely creeping into the Alberta consciousness. Sherman and Orr were rabble-rousing, but there were other contributors amplifying a message that, around the world, and perhaps in the province as well, craft was on the rise.

There is currently no shortage of blogs about Alberta beer. There is, however, a shortage of those that tell the whole story. My own was an excellent example. *One Year of Alberta Beer* was, as the name suggests, by no means comprehensive. It was a snapshot of the industry as seen from the bottom of more than a few pint glasses, and therefore a blurry snapshot. I had actually not felt the need to capture the totality of an industry, given that Jason Foster had already been doing that at onbeer.org, where he's been writing about beer, mostly local, for a decade. Based in Edmonton, Foster made his connection to beer as a homebrewer. He became a beer geek before beer geeks were cool (or considered as such by at least some people), and he made no effort to hide it. Even back in the early 2000s, "I was the obnoxious guy in the pub who asked what beer was on tap and expressed utter disappointment that the best thing they had was Guinness," he told me.

Soon, that love of better beer led to an offer of a platform upon which to explore and celebrate it. In 2006, a friend and editor at *Vue Weekly*, a now-defunct alternative weekly newspaper, made him a proposal: "You know, you're so obnoxious," he told Foster. "We just started a wine column. Would you write a beer column?" In 2008, the gig led to a weekly spot on CBC Radio, as well as beer-related magazine writing.

Arguably, it led most importantly to Foster establishing onbeer.org. Beer blogs were popping up across Canada as craft beer scenes exploded in British Columbia, Ontario, and Quebec. One region was clearly missing.

"I started to realize there was just a big hole in the Prairies." No one else was covering the community, Foster recalled. "So, I said, 'OK, I'll do it.' I could write what I wanted, I could write when I wanted. Alberta had, like, twelve breweries; Saskatchewan had three; Manitoba basically had Fort Garry [Brewing] and Half Pints [Brewing Company]. I thought, 'Well that's not that hard of a job.'"

Onbeer went live in March 2010. Foster posted two or three times each week, beginning with news about craft beer across the Prairies and profiles of new breweries. But it was much more than that as well. What solidified Foster's stature as the "beer guy" (that's @ABbeerguy to his Twitter followers) and distinguished him from other beer bloggers who would emerge in the province, was the academic slant and opinions that he brought to some posts. This made sense. After all, his alter ego is Dr. Foster, an associate professor at Athabasca University, the online degree-granting institution where he specializes in unions and labour relations.

"I'm a bit of a policy wonk," he said.

When various levels of government would announce changes that affected the beer industry, he'd delve into the why and how. This included posts on city zoning bylaws, changes in how craft beer producers were taxed by the province, how the federal government got in on that action, and several posts, naturally, on deregulation. On the whole, this was not often sexy material, nor did he sell it that way, as evident in posts titled "Court Grants Injunctions in Mark-Up Cases," or "Beer Status Quo in New Provincial Trade Agreement."

"I am almost sorry to write this post," Foster wrote in one, "as it dives into the depths of internal trade deals. But that, apparently, is what I do, so here goes . . ."

But policy posts turned out to be some of the site's top attractions, indicating a growing thirst for an understanding of how the industry worked in Alberta. As the blog grew, attracting nine to ten thousand unique visitors each month, it took on a sense of responsibility for Foster. "So, there was some onus on me to sort of . . . hold that up." Today, *Onbeer* remains the clearest and most comprehensive picture of the industry online today.

Looking back on the body of work he's produced thus far, he hoped that it had been impactful. "I know the policy-makers are reading," he said. Just as important to him, however, is the possibility that he may have inspired a homebrewer to scale up, and thereby contributed to the industry we know today.

Foster may not be willing to take much credit for the state of the industry today, but he was chief among promoters, critics, and defenders. He still is, despite a number of influential voices that have joined the discussion post-2013, by virtue of having been a commentator on Alberta's craft beer industry when few people would have considered it an industry at all.

"It's kind of fun to think that, maybe in some tiny, infinitesimal way, I was a part of help making it happen," he said before adding, "I mean, of course, mostly I wasn't."

Homebrewers would be more likely to say that Foster's posts had a positive influence on them than the AGLC would (when asked, sources said that the blogger's posts had not directly affected policies). But it's hard to imagine that, overall, he didn't have an impact.

Sherman and Orr, after all, could only do so much.

BEST BARLEY, BAR NONE

Another possible contributor to the zero-to-one-hundred response following deregulation may have been the one simple fact Sherman and Orr relentlessly highlighted: Alberta has some of the best barley in the world, and plenty of it. I discovered just how much during a trip to a central Alberta village in spring 2019.

At a glance, the village of Alix, population approximately seven hundred, seems like a typical small farming community in central Alberta. From the highway passing through, there are the familiar sites of an old tin-clad arena, a forlorn motel, and, in general, an awful lot of space with not a lot of people around to fill it. But this is no ordinary rural Alberta town. It is a cultural catalyst of a kind. Located about a half-hour's drive east of Red Deer, Alix is the alpha to the omega that is some of North America's finest craft beer. Rahr Malting, one of the largest barley processing facilities in the Americas, anchors the southeast edge of town.

"You'll know it when you see it," Kirk Zembal had told me as we approached in his late-model Subaru station wagon. As one of the co-founders of Blindman Brewing, he enjoys a certain familiarity with Rahr, a forty-minute drive from the brewery in Lacombe, an agrarian centre that has grown into a city of more than thirteen thousand. "We're just down the road from a malt house that ships all over the world."

The road rose and fell over hills covered in the tan corduroy of last season's crops; the rows rigidly precise. Farming had not struck me as such a fastidious profession until I joined Zembal on a few of his annual courtesy calls to suppliers and growers that Blindman could not do without.

Once we'd hit Alix, I could see that he was right: Rahr was unmistakable. Grain elevators, the rural equivalent of skyscrapers, are common sights on the Prairies, but quaint in comparison. Established in 1847 in Wisconsin, with its Alberta plant set up in 1993, Rahr Malting comprises a series of massive metal bins clustered at the foot of a concrete, circular tower that rises to the height of a twenty-two-storey building.

Zembal had been here countless times before. Inside the main office he greeted Bob Sutton, vice-president of sales and logistics, like they were old buddies. He'd come by to pick up a couple of special bags of malt and show me the connection between the vast majority of Alberta's big-city taprooms and the rolling hills that surrounded us, soon to be sown with what ranks as some of the world's best barley. Sutton, now in the middle of his third decade with the plant, showed me a map of the United States. It was black and white, rectangular, about a metre wide and riddled with pins. And while Sutton had been to some of the locations they marked, the pins weren't for him. They were for the barley malt, the heart and soul of beer.

The map was a bucket list for any devotee of North American craft beer. Rahr malt ends up in the beers of Sierra Nevada, Lagunitas, Russian River, New Belgium, and countless other legendary breweries. When Sutton talks about Lagunitas founder Tony Magee, who has visited the area often, he simply refers to him as Tony. Part of the reason for the preference is the quality of the local barley, but another part is how Rahr has honed its processes to ensure that great barley can be turned into great beer. It's a commitment that Sutton, with 85 percent of his revenues coming from craft brewers, takes very seriously.

"The plant operates 365 days a year, 24 hours a day," he said. "There's always a shift on; we don't turn the lights out at night. Some places do, but that's not Rahr's way. Somebody's got to babysit the barley."

All of which means that Alberta brewers have access to the same stuff and therefore the same advantage as the great American craft beers they compete with on liquor store shelves. There's one important difference: Being located a few hours in any direction from Alix, Alberta brewer's get the same stuff, only fresher.

"Start with quality, you should end up with quality," says Sutton.

That quality actually starts well before the maltster. In part, that's an accident of geography. Alberta is a continental powerhouse when it comes to barley production. Not only does the province produce more of it than the rest of Canada combined—more than four million tonnes annually—it produces more than the entire United States. The reason, for the most part, is the climate of central Alberta. Barley matures quickly, which suits the region's shorter growing season. What's more, it thrives in cooler temperatures that help keep diseases at bay.

These factors mean maltsters can get picky, suggested Lori Oatway, a grain quality research scientist at the Field Crop Development Centre in Lacombe. Established in 1978, it follows in the footsteps of the federal Lacombe Research and Development Centre, which came into being in 1907 to help develop the local livestock industry. With so much barley being produced, "we can be very selective about what goes into malting and what goes into the feed industry," Oatway said.

To that end, the centre takes an active role in developing varieties. "We all want to make our farmers successful," said Oatway, who's been with the centre for more than twenty-five years. Barley that goes for malt is worth more than what goes to feed. Not every variety of barley, however, makes for good malt. With a lab in downtown Lacombe and a field just a kilometre south, the centre is dedicated to identifying the cream of the crop. After crossing varieties with others from similar facilities in Saskatchewan and Manitoba, staff at the centre will plant the outcome and analyze it using non-destructive near-infrared testing. By looking at what light passes through the barley and what's reflected back they can determine protein, starch, and fibre content. It's a like an x-ray of potential malting quality. Oatway said they could process as many as forty thousand samples during lab-bound days between fall harvest and spring seeding. When they find promising crosses, they'll send about half a dozen seeds as sacrificial lambs to disease nurseries around the world to see how they fare against scourges like fusarium head blight, a mould that is decimating crops in other regions, including Manitoba. The centre's tests, however, can go much farther afield. Before it ever ends up as a local brewer's malt, a variety may have its mettle tested in Africa or South America.

That's only the beginning of the gauntlet for the grain. Once strains are singled out for quality, yield, and disease resistance, they're entered into a co-op system in which they're grown by farmers across the country and studied. If all goes well, they're submitted to a committee composed of brewers, agronomists, and other industry members. Here, the fate of those varieties is sealed.

"Some of the new varieties perform very well agronomically," said Oatway. "Out in the field they'll look phenomenal."

But brewers can be a fussy bunch. "We'll get feedback that [new varieties] don't have the flavour of the historical varieties." If an evaluation committee doesn't get "very excited" about a variety, she said, "they'll totally wipe [it] out."

Rather than find itself at a dead end in a process that can take as long as fourteen years, the centre has also found a way to hedge its bets. It collaborates with a certain partner located just a few minutes from the office.

"It's very easy to go down to Blindman Brewing and say, 'Hey, we have this variety, we're not sure what the flavour is like. Would you be willing to do a small test batch and give us feedback?'"

Those beers—government experiments that occasionally make Blindman a kind of Area 51 for craft beer—are part of the path to the more sustainable industry that Oatway and her colleagues at the Field Crop Development Centre are working toward. The partnership can help speed up commercialization of varieties that not only have the flavour profiles brewers want, but the resilience to stand up to diseases that are making inroads into regions where they were previously rare.

Between 2001 and 2013, fusarium was turning up more frequently in southern Alberta. It has since begun to encroach northward. While in 2001 just nine counties in central regions reported the fungus, which causes yield losses and affects flavour, twenty-six had identified it in crops by 2016. Faced with climate change that may eliminate the advantage of cooler evenings and the ever-present risk of contamination that comes with the global trade, the centre is in a race against time. It's a hard job, but Oatway is only too happy to do it. Someone, after all, has to help taste those test batches made with malt that may one day be heralded as having preserved an industry.

"It's part of my job description," said Oatway cheekily. "It's very important to get involved with what you're working on."

THE CORPORATE BEDROCK OF CRAFT BEER

Even while Oatway and the other staff at the Centre continue their search for the perfect barley variety, there is no shortage of pretty decent product in the meantime. The irony of Rahr Malting is that although it is the bedrock of craft brewers, many of them producing small batches relative to multinational operations, it is no small-batch outfit. Its output is staggering. The plant pumps out 140,000 tonnes of malt each year, in 480-tonne batches—"biggest batch size in the Americas," Sutton said, proudly.

He has good reason for that pride. It is an extraordinary facility, exceeded in size provincially only by the Canada Malting plant in Calgary (which, incidentally, is very close to the Inglewood rail yard site of the original Calgary Malting and Brewing Company). As we walked from the Rahr offices to the malt house for a look, we'd pass the occasional pile of raw barley fallen from one of the trucks in from the farms. Each was easily enough for a few batches of homebrew—but nothing, of course, compared to what was inside.

On the top floor of the facility, which would have offered a marvellous view of the countryside were there any windows, five pools of grain soaked so as to trick it into thinking it was time to get growing. It moves downward through the tower after that, draining into a larger single pool to start rooting. All the while, the barley's tough husk is being slightly degraded and enzymes required in the brewing process are being called up for service. Then, at just the right time, they're asked to stand down.

"About halfway through the process, the barley becomes malt," said Sutton as we stood in the kilning room at the bottom of the tower. Unlike the room at the top, where the air was still and cool, the kiln was warm and windy. We leaned on a rail and looked upon a circular sheet of barley roughly a metre deep and raked into concentric circles like a Zen garden. Over twenty-four hours, the temperature in the kiln would rise to about 100°C, reducing the water content of the malt from about 45 percent to 4. For now, however, the kernels were soft and pasty. Using a cup on a long rod, Sutton reached in and scooped some out. They tasted more starchy than sweet, and none too beery.

The tasting, however, would improve at our next stop. We left the malt house for a nearby lab where that malt is converted to a simple wort, the sweet liquid that yeast transforms into beer. A technician had a batch already made and sitting in a conical flask on a counter.

"There should be no off-flavours," said Sutton as he poured out small glasses for me and Zembal.

It was perfectly clear and of a colour somewhere between dry straw and mid-summer sunshine. It tasted mildly sweet and vaguely mineral, as if licking a piece of slate, but, compared to what it could become, boring. That's exactly what Sutton was after. If Rahr has done its job right, it should deliver nothing more or less than the expected results each and every time, without fail.

"With base malt," Sutton said, "consistency is next to godliness."

Boring was his benchmark of success. The rest is up to the brewers.

RIDING THE BARLEY WAVE

The craft beer boom will not likely be threatened by a barley shortage in Alberta anytime soon. If anything, more farmers may be enticed to give over acres to growing for malt, given the direction of certain agricultural markets. Scott Keller, a fifth-generation farmer in his late thirties, grows spring wheat, canola, peas, fava beans, and malt barley on twenty-five hundred acres not far from Camrose. He's one of a select group of contract growers for Lagunitas Brewing Company ("Tony's" operation, were you to ask Sutton), a founder of the American craft beer boom.

That relationship is as important to Keller as it is to the brewer. At the time we talked, wheat prices hadn't been great, he said. Also, the Indian market for pulses had been closed for over a year and had kept pea prices down as well. Then, as the result of a trade and diplomatic dispute, the Chinese government banned Canadian canola imports in March 2019, cutting off a market that had helped ensure consistently good prices on the near-ubiquitous Alberta crop.

Barley, to some extent, fills the gap, Keller said. "To have a second crop that has the ability to make you the same kind of money [as canola], that's crucial." And he believes that's not going to go away. In his view, interest in the barley crop for malt rather than feed is only just beginning in Alberta.

Keller, mind you, has a bias—though it took a long time to realize it. Once, his go-to brew was an ice-cold light lager. Then, in 2010, his parents went on a field trip with Rahr to the Sierra Nevada Brewing Company in Chico, California. They brought back some of its signature pale ale, a brilliantly hoppy beer that helped launch the craft beer movement in the US, and so North America, in the early 1980s. It is consistently rated one of the world's best craft beers.

Keller hated it. "I thought, 'Oh my god, they're wrecking our malt,'" the first time he tried it. "'Our barley goes into making something that I can't even drink.' It took me an hour to get through that beer."

But on some level he was hooked. He soon found himself trying other craft beers. "Then, all of a sudden instead of drinking Coors Light, I was drinking Big Rock—Grasshopper or Traditional. It just started snowballing."

In fact, Keller got so keen on craft that his farm abetted a beer to be made entirely of Alberta ingredients. His farm was the one from which Tool Shed's Graham Sherman harvested yeast, collecting it in an open tub of wort driven through Keller's fields. Prairie Pride Ale was released in summer 2017. After that, Keller put in a special request at the local liquor store to stock two of Tool Shed's core beers—Red Rage, a dark ale, and Star Cheek IPA.

Not all farmers share Keller's sense of craft beer fandom, though they still appreciate the fact that malt barley goes for a higher price than when it's sold for animal feed.

"The last few years, it's been so tempting to put more malt in," said Doug Parcels, who tended to plant five to six hundred acres of it a year. Of all his crops—which also included canola, wheat, and peas—malt barley tended to be the most profitable. Parcels has been farming near Red Deer for more than four decades, and he was on Zembal's list of courtesy calls to make after our visit to Rahr. As we'd arrived, Zembal had set two flats of Blindman River Session Ale on the seat of a forklift parked outside of a large grey machine shed. Parcels, his face friendly and slightly wizened, looked at the forty-eight free pint-sized cans of beer more obviously curious than outwardly appreciative as he and Zembal got to chatting.

"So, you get the bag of malt," Parcels said. "Then what do you do with it?"

After Zembal explained the process, the conversation continued in a similar vein, two people working with the same product but in entirely different contexts. For Parcels, the job was seed, feed, reap, repeat. But in talking to Zembal, the process seemed to have found a way to break out of that cycle. Parcels asked him how much beer Blindman makes, where it gets sold. He asked what made their beer "craft." He asked the goals of the business.

"Just ride the wave," Zembal responded.

As the minutes passed and the conversation slowed, Zembal suggested that he should let Parcels get back to it. Before he did, however, the farmer paused. "That's interesting that some of it has stayed that close to home," he said of his barley.

We headed to Zembal's Subaru. As I looked back, I could see Parcels disappearing into his shop. The flats of beer still sat on the forklift.

NEW HOPE FOR LOCAL HOPS

As Tool Shed's Prairie Pride demonstrated, it's now possible to brew a beer made entirely from Alberta ingredients. Of beer's four core ingredients, barley, water, and yeast are relatively easy to come by. Hops are the challenge. Thanks to an agricultural movement spearheaded by sisters Catherine Smith and Karin Smith Fargey, that's changing. In 2011, the sisters established Northern Girls Hopyard at their Windhover Farm, an hour west of Edmonton near Lac Ste. Anne. It was Alberta's first hop crop, and at the time the most northerly on the continent.

"The 'girls' are not Karin and I," Catherine told me in the weeks before the start of the 2019 growing season. "The girls are the hops themselves, because all hops in commercial brewing are female."

Boys, it turns out, don't produce the flower cones that give beer its pleasant bitterness and myriad flavours, ranging from spicy to piney to citrusy to tropical and more. North of about latitude 55, or 2 degrees up from Edmonton, female plants won't do the job either, as flowering is triggered late in summer by diminishing daylight. "It's pushing the envelope in the hops world," said Catherine.

Though both sisters have backgrounds in agriculture, the hopyard started as a "what if." They'd been looking for an interesting addition to their operation, which was anchored by an apple orchard. A phone call from Karin's nephew in Ottawa tipped them off to the burgeoning hops-growing community in the east. Why not here? they thought. They knew it was possible, in theory. Soon, they began cutting larch poles from the farm's natural tree stands and setting up trellises for a handful of varieties chosen for their prevalence in the brewing industry—varieties that local brewers source mainly from the Pacific Northwest, a region that grows roughly 97 percent of all hops in the United States. In 2017, growers there harvested almost 105 million pounds from 53,282 acres comprising dozens of varieties.

Catherine and Karin started with less than one acre. By 2019, that number was unchanged. The difference was that by 2015 those first plants had reached maturity, meaning better yields. Even by 2016, they were harvesting amounts that were close to that of their US counterparts, which were taking in an average of 1,959 pounds per acre.

Back then, though, the goal was only to prove the concept. "Now we know that north-central Alberta can [produce] hops that are early mid- to early maturing," said Catherine.

For brewers, this was a novelty, but also a potential market differentiator. After trialling fresh hops in a cask in September 2014, Edmonton's Alley Kat Brewing was the first craft brewer in the province to commercially release a beer made with locally grown hops. Their Alberta Dragon Double IPA featured Centennial and Cascade varieties from the Northern Girls Hopyard in 2015. As time passed, more brewers took note, such as Tool Shed, but also Blind Enthusiasm in Edmonton, which was keen on making what's known as a wet hop ale using fresh Golding hops from Northern Girls.

"Making a wet hop ale is a huge highlight for any brewery as it's an opportunity to use unbelievably fresh hops," Blind Enthusiasm owner Greg Zeschuk wrote in a blog post about his visit to the hopyard in fall 2018. He helped harvest sixty kilograms of cones. The trick is to move them off the vine and into the brew as fast as possible. Being just over an hour away, he and his team were able to pick in the morning and have them in the mix by the afternoon. "As you may imagine, it smelled incredible," he added.

Such beers, however, are made infrequently and in small quantities—a bit of seasonal fun. Most brewers keep contracts with large hops suppliers to ensure a supply of dried, pelletized hops, which store better than fresh product. "I think one avenue for a small hop producer is in fresh hops, which some Alberta craft brewers [use] in the fall," said Alley Kat's Neil Herbst. Regular usage didn't strike him as viable at the moment.

Just the same, demand has proven large enough to nurture Alberta's hops industry into seedling stage. By spring 2019, there were twenty-four acres under cultivation in the province, said Jason Altmiks, president of the Alberta Hops Producers' Association (which included twenty-one members at the time, accounting for most of the province's growers). That was up from three acres just two years earlier, echoing the local craft beer boom. "Growers want to contribute to the amazing Alberta beer scene and be a part of a product that can source all of its [ingredients from] Alberta agriculture," Altmiks said by email.

What's more, Catherine felt the practice of hops growing in Alberta was finally being given the credit it was due. In 2016, Olds College committed to research varieties and cultivation, and found that 93 percent of Alberta brewers say the need for fresh, local hops is not being sufficiently met. Soon after, Alberta Agriculture also began collecting data on the crop.

"A lot of the building blocks for the industry are in place," said Catherine, who hopes to see her hops acreage double this year.

While Northern Girls already sells out of varieties each year, that kind of institutional validation is key to the industry's sustainability and to strengthening the local supply chain. It moves the industry closer to having a reliable Alberta source of an ingredient that is essential to making unique, world-class beers. For Catherine and Karin, it's one more opportunity to connect consumers to agriculture, and emphasize the importance of producing and consuming local products in support of the provincial economy. More simply put, Catherine would describe it as building community.

For Northern Girls' fifth anniversary, brewers from Alley Kat shared a tasting of their fresh-hops IPA out among the plants. Catherine remembers the event, as "sublime. It's a full-circle moment—a sense of fullness. It's really important to Karin and I to be able to share that."

ALCOHOL AND ALCHEMY

The same December that deregulation came into effect in 2013, Tool Shed received its brewing licence from the AGLC. The event was captured in a tweet Sherman shared, featuring a picture of him and Orr smiling widely as they accepted a manila envelope from an AGLC staff member. They were, appropriately, the first to do so, essentially launching a new beginning—even if only symbolically—for the industry. Not long before my conversation with Sherman in spring 2019, his wife, who works at a dental office, was talking to a client who happened to be starting a brewery in Calgary. She asked him how the process had been going.

"The guy goes, 'Pretty much seamless. Applied for a licence, got it. Applied with the city, got our space. Went to the bank, they lent us all this money. It's been great,'" said Sherman. "My wife's like, 'Oh my god,' having to bite her tongue."

There's been so much growth, Sherman said, that the forgetting has already begun, like a rapid generational turnover between the one that fought in the trenches and the next that enjoyed the freedom it brought. To the guy in the dental chair, starting a brewery was "no big deal," he felt.

"It still would be a big deal if we hadn't got those laws changed."

Some will argue that to say they "got those laws changed" is to say too much. The policy reviews were already underway, they'll tell you. When I put the matter to the AGLC, the representative politely declined to say anything more than that it was a collective voice to which they listened and responded. But Sherman and Orr's story undoubtedly humanized the struggle during the AGLC's deliberations. Being allowed to open their business on their own terms in Alberta would help them, of course, but it also allowed them to make the case that this was about something bigger than Tool Shed. Their story was one ingredient in what they were touting as a possible new chapter in the province's economic development, a story about Alberta and its beer. Of course, at the end of the day, in this case the one I spent making the rounds with Zembal, all the ingredients that have gone into making a great glass of Alberta beer really only become clear when tasting a great glass of Alberta beer.

After a quick tour of the brewery, including a recently installed, very large thirty-barrel brewhouse that qualified Blindman as one of the province's bigger operations, Zembal took me to the taproom to try anything I wanted. It was the equivalent of the candy store scenario and I froze, overwhelmed by about a dozen beers on tap. You pick, please, I told him.

He poured me two glasses of Lemons & Limes, a fruited kettle sour, light and zesty. One of the glasses, however, was nitrogenated, meaning its bubbles are mostly nitrogen gas instead of carbon dioxide. Like a pint of Guinness, that dark and famous Irish stout that owes its creaminess to "nitro," the modified, slightly hazy Lemons & Limes had less of an edge and was possessed of a little more pungency. Though both were excellent, they seemed completely different.

They were the same, however, in one important way: They were so extraordinarily distinct from the crystal-clear, syrupy wort Sutton poured us in the lab. That may seem obvious, as what was in that flask was not beer at all, but its precursor, and in this case the simplest version of that precursor. But the beers Zembal poured for me were fruity, malty, and just sour enough to remind one of the sweet-tart candies of their youth—that is, so unlike the conventional idea of *beer*—that it was difficult to imagine them as having any lineage whatsoever to Rahr's wort.

As a man of science, having taken but never used in earnest a degree in cell biotechnology, I chalked up the effect to one thing: alchemy. The making of a kind of liquid gold had become possible. Local barley was but one ingredient, Alberta-grown hops another. On top of that there was a healthy dose of advocacy and championing, along with a government that felt it no longer had a place in the brewhouses of the province. But there was one other key element that was clear in those glasses of Lemons & Limes.

There was a great deal of talent and creativity just waiting for this moment in the history of Alberta beer to finally arrive.

BEER SCHOOL

A lot of that talent was brewing at Olds College, and almost coincidentally. The brewmaster and brewery operations management program at the central Alberta institute preceded deregulation, with its inaugural intake in fall 2013. According to Peter Johnston-Berresford, the instructor who first proposed the program, neither he nor anyone at the college knew of the legislative changes to come. At the time, there were seven breweries in the province, compared to more than a hundred today. A few grads might find work locally but most would have had to look to British Columbia, Ontario, Quebec, or even the Pacific Northwest, jurisdictions that hosted healthy craft beer industries. A sense of impending change didn't motivate him, as no industry was then clearly on the rise. Instead, the motivation was, at root, all that barley.

Johnston-Berresford's timing was good, as the moment seemed right for the college to expand its offerings. For about a decade, he said, the polytechnic had not added a new program and the executive was open to the idea to "shake things up a bit." As a coordinator of horticultural programs at the time, beer struck Johnston-Berresford as a good way to do that. He was familiar with the brewing program at Ontario's Niagara College, home to a horticultural program from which he had liked to "headhunt" students for Olds. An unabashed straight-talker, Johnston-Berresford does not shy away from telling me how he ultimately decided to table the possibility of the program: "I stole the idea."

He did not, however, expect selling what he'd stolen to be so easy. "The idea of making liquor and calling it a program is like, 'Are you shitting me?'" he said. "No one was sure how it would float with the [college's] board or the Ministry of Education. But, guess what? They were incredibly receptive."

Though there was an apparent lack of a local industry to serve, a brewing program fit the vision of the college, which includes leadership in education specializing in agriculture. "It reflected value adding of an agricultural product—cereals," said Johnston-Berresford. "And it was so original and so different, it placed Olds College at the forefront of new product development."

What's more, it did so in support of one of Alberta's oldest entrepreneurial models: the family farm, the future of which worried the instructor.

"[It's] my belief that if we don't create sustainable conditions in rural Canada, we will lose the family farm," said Johnston-Berresford. "I see brewing as an extension of value adding in terms of what we do in this province. I walk a hundred metres in the spring here and see barley that's going for malting. We're right in the middle of the barley belt. An extension of that is that every single community should have a brewery," he added. "It should be creating jobs that are valuable, creating community-sourced products that are consumed locally. It ties in with the whole notion of self-sufficiency and sustainability."

To Johnston-Berresford, it also ties in to the larger provincial pursuit of economic diversity. Looking at gross domestic product, Alberta hasn't changed much in the last decade. Mining, oil, and gas account for about a quarter of the economic output year after year. During the same time, agriculture and forestry held fast at just under 2 percent – down from 3.6 percent in 1987 (keeping in mind, however, that the provincial economy ballooned almost five-fold during that time, to just under $332 billion). The refining of barley into beer, then, might be a redistribution of eggs, leaving a few less to be broken when the oil and gas basket takes its periodic tumbles.

"Instead of being dependent on one damn commodity, we can start doing things with some of the commodities we tend to ignore, and create value in other ways," said Johnston-Berresford. "It will never equal oil and gas or coal, but [agriculture] creates diversity in and of itself."

Even after its first year, the brewing program began to validate Johnston-Berresford's prediction that it would prove its worth to the provincial economy—and in doing so showed signs that it could become a victim of its own success, though this certainly had something to do with the fact that deregulation occurred around the same time. Of the twenty-six students admitted into the inaugural class in the fall of 2013, five left the program after about a year, recruited by startups eager to be among those first to market in a new era of Alberta craft beer. Those students had yet to secure a diploma, but they'd amassed rare skills and knowledge, making them seem like good hires despite not being graduates. This irked Johnston-Berresford. The program was designed to produce well-rounded denizens of the industry, with an understanding of workplace communications, business math, sales, and promotion, and not just a knowledge of how to work the equipment.

"I can't blame them because of [the money they start earning]," said the instructor, "but they're shorting themselves in terms of learning."

For Garret Haynes, not finishing at Olds College was not an option, even if he'd already made a promising connection with a guy who was making the transition out of his Red Deer garage and into a bona fide brewery he planned to call Troubled Monk. Brewing was like a calling for Haynes, and he wasn't about to do anything that might compromise his chances now. Before school, he'd hit "a wall" trying to break into the business, he said. "I would just give out resumés and talk to people and nothing would happen."

Were you a brewery owner at the time, you might not have believed Haynes had much going for him either, even if he was a long-time homebrewer.

As Johnston-Berresford puts it: "Homebrewing is great but it's a bit delusional for people to think, 'If I can make a great beer, I can make money doing this.' There's a shit-ton more information and experience you need before you even contemplate going into industry."

(Call him biased if you like. If so, try the homemade beer on offer at your next backyard barbecue. Maybe you'll find yourself in the presence of a brewer who knows his or her stuff, maybe you won't. Good luck.)

After growing up and graduating from high school in Ponoka, a town of seven thousand people forty minutes north of Red Deer, Haynes did not immediately set to accruing that "shit-ton" of information and experience. Instead, he completed a digital arts diploma at Red Deer College and later tacked on a fine arts degree at the University of Lethbridge. The result was a vagabond's LinkedIn profile: Haynes has variously made a living by working in an auto body shop, testing video games, construction, and videography. He even ran his own business designing and making T-shirts (he's since put those skills to use designing the art on cans and other collateral at Troubled Monk). While wandering from one position to the next, he did a couple of stints of wandering the world. That's what set him straight and put him on the path to craft beer.

"I was a homebrewer for just under ten years before it occurred to me, looking at my resumé and all the jobs that I had [done] and not loved, the one thing that I'd been doing and did love was brewing beer."

After his series of resumé rejections, a friend heard on the radio about the brewing program starting at Olds College. "I made up my mind to get serious about it and go to school," said Haynes.

As he progressed through the program, he delved into the technical side of the industry: the chemistry and biology of making beer, of which the casual home-brewer tends to have but a cursory understanding. Haynes would also end up well prepared for his career, ironically, by some of the program's shortcomings. His class was equipped with a brand-new, never-before-used brewery. The tweaking and adjusting and repairing that followed provided a series of teachable moments.

"Every time something went wrong, I was taking notes. And that translated directly in going to Troubled Monk. I knew a lot of problems that we were going to have. So, I think that our class is definitely luckier than any other group." A few days before I'd met him, for instance, he was replacing bearings in the canner. "That's what being a brewer is."

Nothing, however, prepared him for the backbreaking labour.

This is an aspect of brewing that Johnston-Berresford spends "considerable time" drilling into students. He feels that he'd be doing them a disservice otherwise. It has been Johnston-Berresford's experience that some prospective students look forward to the slim possibility of attention and accolades that can come of the job and discount the guarantee of aches and pains.

"I keep saying, 'Brewing is not a fucking Twitter post—brewing is hard work,'" said the instructor. "Some people can't hack it. They realized like, 'Holy god, my feet are soaking wet half the day.'" Johnston-Berresford sees the school as a filter, eliminating wannabes who "don't have the calling." For some students, "it's a bit of a shock to them how blue collar it really is."

Haynes would agree—or at least his body certainly would. "I dropped about twenty kilograms in the first couple of months of working at Troubled Monk," he said.

When the brewery opened, with Haynes as head (and only) brewer, it did so with none of the conveniences it has today, like the grain hopper that now sits at the back of its parking lot, or even a forklift. Back then, "artisanal" truly meant that every step was completed by hand, like it or not. Malt arrived in twenty-five-kilogram bags on a pallet on the back of a pickup truck and the sweat-inducing, pound-shedding labour would begin. Haynes would join a crew including brewery co-founders Charlie and Graeme Bredo in "hand-bombing" the bags to another pallet on the ground. Tacked onto actual brewing, manual labour made for long days. "I don't know if I had more than one or two eight-hour days in my first year and a half." Some days ended at two in the morning. In fact, only in the few months before I met Haynes had life settled down enough, and into reasonable hours, that he'd been able to commit to relationships with people other than his co-workers, and recently got engaged.

Haynes's introduction to the realities of running a brewery are not unique. "I got my ass kicked when I first started," Tracy Vornback told me. The 2018 Olds College grad was a brewer at Tool Shed when I spoke with her. "I was definitely not prepared for how physical the job is, especially during the summer months and when the brewery is thirty-plus degrees."

Vornback came into the program when some of the bugs Haynes encountered with the equipment had been worked out, missing out on that golden age of impromptu tinkering and problem solving. "School educated me on things like the biology of yeast, but diagnosing a pump failure is all hands-on learning as you go," she said.

One might think that post-secondary would be the perfect setting to also develop a high tolerance for stress. No matter where or what a student studies, the system almost always seems to court breaking points. Coursework takes up the days, homework eats up the evenings, and a job steals from both, and often sleep as well. It's up to the student to find ways to hold it together, not to mention develop a taste for budget beer.

Life as a brewer, it turns out, is an extension of that mindset (though the beer ought to be better). As manager of the Olds College production brewery from 2014 to 2017, Dave Mozel, the current brewmaster at

Folding Mountain Brewing, in Hinton, did his best to prepare students for the realities they'd encounter in their careers.

"I tried to give them a feel for what the real industry was like—not just sitting in a classroom," he said. Part of that included fostering mental toughness. "We have a lot on the line when we have a tank of beer going. You have to be able to handle that. You have to learn to chill out, let the beer do its thing. I guess we're kind of laid back most of the time and highly stressed out at other times. You're waiting three weeks for that beer to be ready and if something goes wrong in those three weeks, you're starting over again."

Looking back on the job at Olds College, he said, "I didn't want to scare them too bad but I'm sure I did scare a few."

What Mozel really wanted was for his students to understand something that might both set them at ease and, perhaps, preserve some humility. "You're not going to be brewing award-winning beer your first day," said Mozel. Then he added with a laugh, "Unless you're Garret."

The story of Garret Haynes's early success, and therefore Troubled Monk's, has taken on a somewhat legendary quality in the industry. All the wet feet and hand-bombing paid off in a way that very few brewers ever enjoy. Less than a year after pouring their first pint on June 12, 2015, Troubled Monk took home a silver medal at the 2016 World Beer Cup, held in

Philadelphia. In North America, if not the world, the international biennial competition, sometimes referred to as the Olympics of beer, is as good as it gets.

"It was quite a night," Haynes told me. "It was just a really floaty feeling walking up to the stage."

He was collecting the award for Open Road American Brown Ale, a dark and hoppy mainstay in the brewery's lineup, and a product of Haynes's talent and knack for dialling in recipes. A fan of brown ales, he's picky about them, often finding them lacking in one way or another. In creating Open Road, Haynes decided to go down the path of the homebrewer and cater less to what the market seemed to prefer and simply brew to his own tastes. He conjured up a test batch but found it too missed the mark.

"The malt had a hole," he said, his tone a mix of an artist trying to put a finger on the ineffable and a craftsperson stumped by a technical challenge. "It didn't have the whole picture that I wanted." So he went about the brewery chewing on different malts to fill the gap. When he found it, he brewed again, and brought the result to the owners. "Everyone loved it," said Haynes.

In fact, they loved it so much, they decided to put it in the running along with the 6,595 beers entered into the awards program.

"He's brilliant when it comes to flavours and combinations and understanding what makes a good beer," said Charlie. "His intuition about what's going to taste good is tremendous."

With Haynes leading recipe development, the team has won several Alberta Beer and Canadian Brewing Awards since. Awards aside, hiring Haynes, Charlie later told me, "was probably the single best decision that we made as a brewery to date." Looking back, the Troubled Monk co-founder admits that he himself was an acceptable though not great brewer. When he placed that help wanted ad to the program at Olds, he learned what he was missing.

IN TRYING TO assess the future based on the contribution of Olds College, Dave Mozel did some quick math for me. He tallied an influx of more than one hundred people trained in brewing, as well as the marketing and sales required to make brewing a business. Not every one of Alberta's breweries had an Olds College grad on staff, but Mozel felt that enough of them did—half, he guessed—to make a positive impact. Among them, he pointed out, were Troubled Monk and other celebrated breweries that had also hired from those first few graduating classes, people like Warren Misik at Edson's Apex Predator Brewing and Derek Waghray, of Calgary's Dandy Brewing Company.

Without Olds grads, Mozel said, "I think Alberta beer would be worse off. There would be a lot less quality. I think the pace of the explosion of breweries would have been slower. I think Olds College definitely had something to do with what happened here."

At the same time, he, like Johnston-Berresford, recognizes that it is only foundational, even if a cornerstone, to the success of the industry. Students inevitably need to build on their skills once they leave the relatively comfortable environment of the teaching brewery.

"You can only see so much in a classroom," said Mozel. He recalled the challenge and frustration of trying to instruct a group of fifteen to twenty students on the school's tiny brewing system. "It's tough to work with one student and show them stuff when you're tripping over another student."

For the educating they can't do in the classroom, the college has struck "a really happy sort of arrangement with the industry," says Johnston-Berresford. Today, the program is structured to allow students to spend a few days a week working in established breweries (which tend to hire them afterwards, anyway).

"It gives some experience we can never hope to give them because we don't have the technology and we don't have the capacity," he said. "We don't even have a big enough brewhouse. Our brewhouse is five barrels [nearly six hundred litres]. I mean, some of these kids are brewing on twenty-, thirty-fucking-[hectolitre] brewhouses—huge compared to what we've got."

That relationship is essential, says Village Brewery's Larry Kerwin, who is chair of the program's committee of industry advisers. Kerwin has hired many interns from Olds College. "[Students] come out pretty well rounded to take on any position that might be available in the industry," he says. After that, there's inevitably learning to be done no matter how thorough an education they received. "From there, they're going to have to evolve. That gives them a good foundation to start. The rest they'll have to learn on the job like any other job."

Tracy Vornback also recognizes the introductory nature of the program, but, after working at Tool Shed during her diploma studies, and subsequently being hired by the company, she also seemed comfortable with what the program could and could not do. "I feel like it gave me a base from which to go on, but gaining work experience as a summer student was invaluable. [It] provided me with solid, hands-on knowledge that is otherwise hard to gain at school."

To its credit, that knowledge has proven enough in some cases. Caswell Johnstone, another 2018 graduate, wasn't able to take advantage of that off-site learning. Johnstone juggled school with raising four children at home in the town of Three Hills, a forty-five-minute commute to school each day.

"The nature of a two-year program is that we simply do not have the time to cover every single useful aspect in depth, or to even have enough brewing time to be considered fully competent on a commercial brewing system," he said, almost alarmingly. Nevertheless, he ultimately found a place as the brewer at

Sawback Brewing Co., a taproom-focused brewery in Red Deer.

"As I began brewing at Sawback, I immediately recognized that I had some weakness in the area of being able to physically walk through the commercial brewing process." Having suspected that might be the case before he got to industry, "I focused on learning as much theory as I could, and as thoroughly as I could, in order to have a solid working understanding of the *why* behind the *what* [of brewing]." What he didn't know right away about the equipment, he made up for in everything from recipe formulation mathematics to yeast microbiology to sensory analysis and more. It paid off. At the 2019 Alberta Brewing Awards, Sawback was part of a three-way tie for the title of best new brewery. But perhaps the most important lesson he learned at Olds about being a brewer, Johnstone said, was to simply "take it seriously."

Part of that involves understanding that making excellent beer can require openness and humility. Johnston-Berresford knows his students "probably won't be great brewers when they leave [Olds College]," not unlike any other kind of student who makes the transition from the womb of post-secondary into the harshly lit reality of industry. "But they at least have the understanding to know what to do with a beer that might be problematic," he said. They're not expected to be virtuosos; they're expected to be problem solvers above all else, and to approach their field for "the trade" that Johnston-Berresford pointedly calls it. They're expected to get their feet wet and be all right with it. The instructor is absolutely intolerant of prima donna tendencies—especially if they interfere with the practical aspects of creating a top-notch beer.

One day at the brewery a student shared a beer he'd made. It was well brewed, the product of a rising talent—the kind of thing that the instructors would be happy to sell in the school's adjoining taproom. But before he tasted it, Johnston-Berresford, whose own teaching focuses on packaging, paused to look at the bottle. Where there should have been roughly an inch of headspace—the unfilled portion beneath the cap— "there must have been fucking four or five inches," he recalled. "I joked around with it, but it kind of irritated me. I said, 'You're shittin' me, right?' I mean, we'd just talked about that stuff.

"I'm not a top-end brewer by any stretch of the imagination," said Johnston-Berresford. "But one thing I'm going to whine about is when [a brewer] can't even pack something properly. We're talking about packaging, but we're [also] talking about the student who's the package."

Johnston-Berresford knew the criticism may have injured the student's ego. He also knew that the student would watch his headspace in every bottle to come.

Over time, that may be what Alberta craft beer needs most: lessons that lead not just to flavours that are unforgettable and unobtainable anywhere else, but the feeling that such quality carries through every aspect of the product. That may mean something as simple as properly filling a bottle, because that's what multinational brewers do with near-unerring consistency. Therefore, that's what customers, and potential craft beer converts, will expect.

To make beers of choice, Johnston-Berresford knows Alberta's next generation of brewers must continue to take at least one fundamental and very basic teaching to heart.

"Have some respect for the beer."

3

SOMETHING'S BREWING

*An industry begins
to simmer*

As 2014 dawned across the Alberta beer landscape, the elements were settling into place: the world's best barley, a deregulated brewing climate, a new hop industry, and a school providing solid training, not to mention a province full of enthusiastic beer drinkers ready and willing to support local craft. All that was needed was people to step up and start brewing.

But such things are never easy, especially when it's all new to everyone, brewers and regulators alike. By way of example, in 2013 a brewer had to produce at least 500,000 hectolitres. In 2014, a brewer could open up shop making literally as little beer as he or she liked. One brewer I spoke to started with so little that when federal officers visited the brewery to measure the output for excise duties, they found that none could be charged at the time. For beer, they're levied per hectolitre, or every one hundred litres, and the brewery hadn't even been able to make that much.

Similarly, Dog Island Brewing in Slave Lake, roughly three hours north of Edmonton, straddled the line between hobby and business for months after owners Ben Fiddler and Chad Paulson started up in October 2016. Within four hours of opening, their tiny tanks had run dry, and they closed up shop. (The company made its name early on the strength of a subtle and refreshing raspberry ale inspired by Fiddler's time spent studying at Brewsters, makers of River City Raspberry Ale, when he was completing apprenticeship studies in Edmonton.) As both Fiddler and Paulson were doing double duty as tradespersons at the time, subsequent brew days were scheduled as shifts permitted. Local demand, however, sustained their stop-and-go business model until the pair secured capital to scale up to a facility capable of satisfying the demand for craft beer in Slave Lake, population 6,650, and beyond.

Still, even if breweries could bide their time, startup was not a simple thing. Immediately post-2013, there was a lag in comparison to the surge that would come in 2016, when roughly twenty breweries would open, and '17 and '18, which would usher in more than fifty. Among the earliest post-2013 entrants are breweries that have since become pillars of the industry in terms of size and quality, including Apex Predator Brewing, Blindman Brewing, the Dandy Brewing Company, Troubled Monk Brewery, and Tool Shed Brewing. To get a sense of just how tough the going could get, even in the enlightened era of craft brewing in Alberta, it's worth taking another look at that brewery started by Graham Sherman and Jeff Orr.

The irony with Tool Shed is that the brewery was never meant to start small anyway. Tool Shed opened as a fifteen-thousand-square-foot facility in an industrial park in the city's northeast. "It scares the shit out of you trying to get a lease on that," Sherman told me.

Their business plan was in keeping with the way they'd approached their military contracts. The client would make a request and Sherman and Orr would push them to extrapolate: Sure, you want this now, but what will your future needs be? With that knowledge, the pair would lay a foundation to suit the long-range forecast as much as the conditions of the day.

For Tool Shed, that almost didn't work. "Life essentially calls your bluff," Sherman told me during a visit I made to the brewery in 2016. "'All right, asshole. You want to make beer the rest of your life?'" he said with a laugh. But it was only funny in retrospect, and then only questionably so. After an initial injection that was reported to be $300,000 to get brewing, the project struggled.

"We were going to go out of business if we didn't come up with half a million dollars," said Sherman. He remembered waking up in tears one night. "I was ruining the lives of everyone around me. I cried quietly so I didn't wake my wife up."

But they'd come too far to fail, and the adversity strengthened their resolve. That same night, Sherman had an idea. He'd offer beer for life: $5,000 for one of a hundred growlers that could regularly be filled at the brewery free of charge, forever. ATB Financial bought one, he said, and so did other breweries, including Big Rock, and Last Best, a Calgary brewpub that had opened the same year, in 2014. The scheme saved them. "These are our competitors and yet we all support each other," Sherman said.

In 2016, Tool Shed made a million litres of beer. Since then, it has leased another bay, eight thousand square feet, making them one of Calgary's biggest breweries. Sherman sees that as validation of not just their vision but of having gone through all that they did to realize it. For him, it was a lesson in life as much as one in business.

"You gotta teach your kids to go after shit in life," he told me. "I tell them to follow their dreams. I don't want them to be the status quo. I want them to go out and create things—accomplish something great and start a legacy. I've got to lead by example."

In a way, that would apply to the brewers who'd lined up for licences behind him and Orr as well. Not many would follow Tool Shed's example, but they would follow the trail it left in its wake. For a while, at least. But, not far along, they'd diverge and find their own way.

YOUTH AND BREWING

Just as you have to go out of your way to visit Tool Shed, visiting Bent Stick Brewing can also require diligence and determination. It's tucked into a tidy and overwhelmingly beige business park in a part of Edmonton (also the northeast, incidentally) that has been undergoing revitalization since at least 1995, when the last of the local meat-packing plants were demolished. Success for that initiative has so far been elusive, with some projects stalled mid-construction. Fort Road, as the area is known, is still a place waiting to happen. Among its biggest draws are a bingo hall and a casino that brings in a steady stream of legacy rock acts. Neither ever give me much incentive to make the trip across the city from my home in the west end. But as Edmonton's first "nano-brewery," Bent Stick does.

Before I set out for it on a muggy Friday afternoon one summer of 2018, I checked Twitter to see what was pouring in the taproom. A tweet less than a day old listed B.S. Wit, a Belgian-style wheat ale, bready and citrusy, with a bit of spiciness: black pepper and coriander. That sounded refreshing enough, but the other offering of Electric Boogaloo, a hazy IPA, promising malt, fruit, and pine, got me charged up. An additional "test-tap" added an element of intrigue. So I hit the road—idiotically, just before 4 PM, when several thousand drivers were tightening a Gordian knot of rush-hour traffic.

Upon arrival, I discovered that it was time for Bent Stick to update its social media feed.

The supply had nearly run dry. The IPA was gone, and there was not even a mention of the test batch (whether it had passed or failed, who knew?). Even the little fridge in the corner of the taproom looked bereft through its glass door, stocked with a dozen bottles at best. All that remained was the wit, which brewer Ben Rix offered when he temporarily left his brewing duties to greet me at the door. "It's one of my favourites," he assured me as he poured a clear, golden pint from a tap just inside the brewhouse. In a moment, all was forgiven. The wit was crisp and cutting, the perfect antidote to twenty minutes of bumper gazing.

As much as I liked the wit, I still asked after the IPA. "This won't happen again," said Rix, talking about Bent Stick's diminished supplies. "We just brought a new tank online. It'll boost production by 50 percent."

I took a seat along a short plywood wall that divides Bent Stick's taproom, an afterthought of a space that is big enough to accommodate maybe six or seven people comfortably, from the brewery proper, a grey warehouse-style space about the size of a racquetball court. I set my beer on a narrow ledge attached near the top of the wall and looked at the half-dozen stainless-steel tanks on the other side.

"Fifty percent sounds like a lot," I said.

"Want to have a look?" Rix asked. With that he began a tour of the brewery, which, given the fact that it is among Alberta's smallest, didn't take long.

Nevertheless, it was a revelation. What Rix showed me, I realized, was a perfect example of how young entrepreneurs are making this industry their own. Manufacturing, which is what brewing is, didn't strike me as a business that might appeal to millennials, a generation typically associated with the unfortunate capriciousness of the gig economy. Set that stereotype aside, however, and a relationship between youth and entrepreneurship and brewing makes good sense, if you consider a few key factors.

One is that entrepreneurship generally tends to be a young person's game—a trend that was highlighted in *Demographics and Entrepreneurship: Mitigating the Effects of an Aging Population*, a book released by the British Columbia–based Fraser Institute in 2018. Age wasn't noted as the only driver, but it didn't take a back seat either.

"The odds of becoming an entrepreneur rise as one moves out of childhood through early adulthood, then fall for the remainder of one's life," wrote one of the book's authors, Russell Sobel, a professor of economics and entrepreneurship at the Baker School of Business, in South Carolina. Generally, he added, those between the ages of thirty and forty-four are most likely to follow through on attempts to become their own bosses. One reason for that is a naturally higher tolerance for risk among youth, wrote Sobel. The older people get, the more they seek financial stability, which isn't a defining characteristic of startup culture. What's more, creativity, another key factor in entrepreneurship, follows risk tolerance right out the door as most of us get on in years. After all, the professor added, "creativity requires taking risks."

The thing that really brings all this to bear on Alberta craft beer, however, is what Sobel himself refers to as an obvious factor: There has to be an opportunity. The spark of inspiration that leads to action still needs oxygen to come alive as a flame. Deregulation, meaning opportunity, in late-2013 was the oxygen. It opened what may have once seemed a closed market to those with a willingness to take a risk and an interest in being creative, and it did it at a time when youth unemployment in Alberta was more than 10 percent. So why not take a chance on a market that ranks as the fourth-largest consumer of beer in Canada, at around seventy-six litres a year? What's more, why not take a chance on an industry that embraces the iterative nature of shoestring startup culture and weaves it into its core values? If a certain beer wasn't well received, it can be taken that it was a good thing that only so much of it was made and it'll be tweaked on the next go. Or, it might be abandoned. That's OK, because customers crave variety. If the IPA runs out because of a lack of tank space, there's still the wit, and that's OK, too, because the shortfall speaks to the nature of hand-crafted products, the unhurried artisanal touch that customers expect of small businesses. The result of young people taking that chance is a kind of symbiosis, in which individual and industry survive because of each other, and even thrive.

When Rix and partners Scott Kendall, Patrick Gaudet, and Kurtis Jensen officially opened Bent Stick in July 2016, they ranged in age from the late twenties to the late thirties, more or less in the entrepreneurial sweet spot described by Professor Sobel. They were well acquainted with risk. At the time of my afternoon visit, Bent Stick was able to brew four hundred litres at a time. It's a lot more than the five-gallon buckets they once tended as homebrewers but it's modest when you account for the time it takes to bring a batch to market. Brewing may only take a day, but fermentation (when yeast turns what's basically grain juice infused with hops into actual beer) and aging (when the flavours mature) takes much longer. Payday is always at least a few weeks away from the moment a brewing kettle fires up. The stress that comes from the uncertainty that underlies production is just a cost of doing business.

Speaking with me during a later visit to the brewery, Kendall described the opportunity that came with the legislation change as having two sides. One, less positive, was likening it to a gold rush. "A few people get rich and a whole bunch of people get shot for their claim," said Kendall, who looked that day a little like a prospector himself, tall and lanky, a thick moustache reaching to the jaw on either side of his chin. "There are victims."

But that's not how the Bent Stick team saw things when they started brewing test batches in their garages in 2014 and '15. Back then, the industry was new, untapped and full of promise. That's why Kendall counters his gold-rush metaphor with a gentler, less avaricious one. The government had "uncorked the bottle and let the genie out," he said. Wishes were about to be granted. Nevertheless, they weren't going to come cheap.

THE ROOTS OF Bent Stick stretch across the city to Alley Kat Brewing Company, opened in 1995 in a south side industrial park. Bent Stick's founding partners put in a few overlapping years as employees at Alley Kat before forming their own company. Kendall looks back on the time as an impromptu brewery school, where they learned everything from how to make beer, package it, get it into restaurants and liquor stores, and more. In addition, Gaudet spent time working in both the retail and distributing side of the Alberta liquor business, and Rix had a certificate from a British brewing institute.

By the time the legislation change came, none of them worked at Alley Kat or in brewing anymore, though they had remained friends. In 2014, Jensen and Gaudet suggested that the four of them brew up a business plan. They talked Kendall into the idea at Brewsters's south Edmonton location, then pulled Rix into the mix.

"We knew we could make good beer together," said Kendall. "We didn't know if we could run a brewery together." They wanted to find out, so they gathered roughly $500,000 between them, including savings and what they could coax from friends and family. They sought out no investors for fear of losing creative control. "We were golden from there."

In using the word *golden*, Kendall spoke in relative terms. He's come to love what he does almost without question; when failure threatened, that love clearly guided him through. If Bent Stick was golden, it dodged bullets to protect its claim.

Another valuable skill Bent Stick owners learned from Alley Kat was how to live within one's means. Early equipment at their former employer was rustic and unautomated. Back then, Alley Kat used tanks with tops that flipped up like massive lidded beer steins. Staff would harvest yeast for future batches by scooping it out with a bucket on a string. Then they'd climb inside the vessel and scrub it down by hand. It was a lesson in resourcefulness that stuck. Bent Stick's vessels are modern, sealed and shiny stainless steel;

they've never had a need to crawl inside one. That said, they're different from most new tanks currently being purchased in Alberta because they're sourced from a Chinese supplier they knew about through Alley Kat. Not only were they cheaper but the wait time was shorter at a time when there was high demand for North American–made machinery.

"We bought the minimum amount of equipment we could to successfully operate and manufacture quality beer," said Kendall. "And we managed by the skin of our teeth to get to market and make it happen."

When they launched their first two varieties in July 2016, the beer was indeed a product of making do. A pale ale called Swap the Hops was, like the tanks, a clever solution to the problem of limited supply. To ensure access to key ingredients, particularly the hops that give beer bitterness, flavour, and aroma, brewers can sign contracts with suppliers. This costs a few thousand dollars up front. Alternatively, they can take their chances and spend only what's necessary to get whatever is available. In that case they take what they can get from one order to the next. Hence, the "swap." While some might see that as a gamble, for Bent Stick it lessened the strain on a chequing account that at times could dwindle to $100. For beer geeks, the brewery's early target market, it signalled a sense of adventurousness, a willingness to experiment that would keep customers guessing. It was in many ways the ethos of craft beer at its purest. "Swap the

Hops was how we made the best of a bad situation," said Kendall.

They still make the beer today. From one iteration to the next, the hops varieties have never been the same. Ironically, that inconsistency is one of the few constants linking the Bent Stick of today to the brewery the partners envisioned at the outset. Much changed over its first year of existence as the owners matured as entrepreneurs. Those changes came to include the makeup of the ownership team.

"If you make the best beer in the world [but] you can't run a financially sound business, it doesn't matter," said Kendall.

A series of pivots that could be construed as philosophical shifts occurred since selling that first bottle of Swap the Hops. "We were just scraping by and making ends meet," said Kendall. He joked that they have a grand total of five lights to keep on at the brewery, but he knew the risk of them going dark was real.

Among those shifts was a change in an approach to distribution that had allowed them to make their debut in Edmonton's craft beer community. At first, they'd refused to sign on with the distributor that puts liquor in stores across Alberta. They were adamant about hand-delivering their bottle-conditioned beer to the restaurants, bars, and shops, many of them the same ones they'd connected with when working at Alley Kat. Jensen, in particular, liked that they had control over where the beer was sold, rather than leave that decision to a third party. But they soon realized that they couldn't work full-time jobs elsewhere, as each of them did, and moonlight as delivery boys in addition to brewery owners.

As the finite nature of time and energy became clear to them, they chose to recoup as much of both as possible by offloading the job of getting their beer out into the world, and signed on with the province's distributor, Connect Logistics. Though it would lack their personal touch, they knew it would help them reach more customers. The decision was made by a vote, with Jensen being the lone dissenter. "After the vote, I accepted the outcome and would help prepare orders to be sent out to Connect," he said.

Expanding the customer base was a more significant change of mindset than it might seem to the outside observer. In the beginning the partners had seen Bent Stick as a connoisseur's brewery, with "beer geeks like ourselves" as the primary customer, said Kendall. As Jensen once commented: "We started as homebrewers. As homebrewers, you just brew the beer you want . . . We're trying to brew beer for us and then sell the [rest] to somebody else."

Kendall wasn't quite ready to relinquish the romantic notion of a "cult following" that can come of that kind of approach to business, but financials have forced him to accept that Alberta is no beer mecca on the order of Oregon, where it's estimated that more than half of all beer consumed is craft, or Colorado, whose craft beer industry is worth $3 billion a year. When it comes to beer on the Prairies, the geek has yet to inherit the earth.

"You need to have a wide appeal to your beers," said Kendall. "You can't just rely on that small percentage."

But there was a more defining act in the maturing of Bent Stick: the adoption of growlers. The idea and practice of take-out beer from the local brewery dates back to the late 1800s, when the vessel of choice was a metal bucket with a loose lid that was said to "growl" as carbon dioxide escaped from the beer. It would take another century until a Wyoming brewer would silk-screen his logo onto a reusable glass jug—a modern growler—and create one of the most iconic symbols of the craft beer industry's independence and distinctiveness compared to multinational brewers. Thousands of companies would follow suit across the continent. Not Bent Stick. At least, not at first.

The argument for resisting was sound. Growlers can be one of the worst ways for a craft brewery to represent itself, said Kendall. For one thing, beer goes slightly flat as it comes out of a tap. That's not an issue if it's streaming into a glass immediately destined for a customer's lips. But when that jug of beer is left undrunk for any period of time the quality inevitably deteriorates even if it's sealed. The other problem is the cleanliness of the reusable jugs, which customers may not have maintained to a brewer's rigorous standards. Almost as bad, it may have been cleaned with soapy water and poorly rinsed. In either case, when the quality becomes questionable, blame tends to fall on the brewer, justified or not. So the brewery decided to start off using single-use capped bottles only.

As time passed, it became clear that a revenue stream was going untapped. When the team finally acknowledged that customers liked and wanted growlers, and that it was time to get silk-screening and filling jugs to go, the decision represented one more in a series of compromises. That did not sit well with Jensen, says Kendall.

Jensen could accept growlers for beers the brewery planned to keg only but otherwise didn't see the point of taking the risk on quality control. It struck him as a time-suck, too. "There is also the man-hours to fill the keg, then to [use it to] fill the growler," said Jensen. "While filling a growler, you can't really help other customers and it takes quite some time, so your throughput is lower [and] slower."

Be it growlers, distribution, or Jensen's concern about investing time in small batches of what he saw as the "overplayed" bourbon barrel–aged beers (one of which, a Bent Stick stout called 11:59 PM, won a bronze at the 2018 Alberta Beer Awards), the gap between him and the other partners grew with the business disagreements.

"I didn't see eye to eye with the rest of the guys and we were having trouble getting together to work past it," said Jensen.

In spring 2017, matters came to a head, and he was asked to leave.

"We decided we needed to stop working together," said Kendall. The difficulty of the decision was compounded by friendship. "That was fuckin' hard for all of us. It's like a divorce. There's so much common history and shared passion, but it's not going to work. That was another one of those business hurdles that test your mettle."

Jensen is no longer involved in the Edmonton brewing community, but declined to say what he was doing now. "I've had some time to analyze and process it and have come to accept it," he said of the split. "I have come to peace with it."

None of the adjustments to the business plan, including parting with Jensen, were ever about becoming Alberta's next big craft brewer. They were the growing pains of new entrepreneurs who were unwilling to walk away from the roles they'd taken on as leaders in a new era of the province's craft beer industry. In a more granular sense, they were on-the-fly adjustments to make sure they could stay in the game, and play according to a plan that, though different, still fit the team.

"What we thought the brewery would be like before we had it was one thing," said Kendall. "The reality of running a brewery was another." The strain of moving money across accounts to cover shortfalls, maxing out credit cards, and draining savings, as Kendall did when he committed to the brewery full time in January 2017, not drawing a paycheque until six months later, had threatened to rob the venture of its fizz. Only a more measured, conscious approach to entrepreneurship would help keep it fresh. Today, Kendall's advice to those considering the craft beer business is unvarnished, coloured instead by frontline experience.

"Don't be as undercapitalized as we were. I wouldn't start with less than a million bucks."

As Bent Stick settled into its third year in business, that kind of frank talk also helped to shape the local industry in the hands of other young entrepreneurs. Sea Change Brewing Co. entered the Edmonton craft beer scene in December 2017 with a simple, well-received blonde ale that was at first available only on draught. Arguably, theirs was a smoother debut than that of Bent Stick, thanks in part to the latter having blazed a trail, however rough and meandering it may have been.

Prior to starting Sea Change, brewer and co-owner Taylor Falk met the Bent Stick team in Edmonton at a craft beer event known as the Real Ale Festival. Organized by Edmonton Beer Geeks Anonymous, it brings together dozens of brewers each summer to tap one-of-a-kind casks of beer for more than five hundred attendees. It's a celebration of small-batch, cask-conditioned beer that, without veering too far toward bacchanalian, resembles a symposium between industry veterans, recent entrants, and those wondering how to get in on the craft beer boom. There, Falk met

Gaudet, who would act as a consultant for what would become Sea Change Brewing, comprising twenty- and thirtysomethings much like Bent Stick.

Taylor needed no help as a brewer. At the time, he was in charge of the tanks at Yellowhead Brewery in downtown Edmonton. But he and his co-owners had questions about how to build a brewing company, even one as small as Bent Stick.

"Patrick's advice saved us a lot of time and money," said Falk in the weeks leading up to the official opening of Sea Change in August 2018. "He helped us decide what we needed to buy now and what we could wait to buy later. A lot of breweries start off too big, without having a track record to justify the purchase of large-capacity brewing equipment."

Falk sees Bent Stick as a model that proves size doesn't necessarily matter. "[Their] growth as a company shows that knowledge and experience are your best assets as a new brewery. There are a lot of breweries in North America with lots of money behind them that are making mediocre beer. Bent Stick proved that you don't need a massive initial investment to make great beer if you know what you're doing. People recognize value and authenticity—especially craft beer drinkers—and I think Bent Stick is a great local example of an authentic, grassroots approach to small business."

Kendall claimed that he's never wanted Bent Stick to become anything else. In 2017, the company managed to make a little more than 211 hectolitres, or

21,100 litres, and he seemed fine with that (Big Rock, by comparison, might make two hundred times that per quarter). He even likes the idea of the brewery being somewhat on the margins of the city, and a bastion of craft in a part of Edmonton where few would think to look for it.

"We're just getting to that sustainable level now," he said. In 2018, they managed 400 hectolitres. The next year, they'd done 285 by July, along with another 200 produced off-site by a contract brewer (which helped meet demand for the B.S. Wit and Electric Boogaloo IPA, incidentally). "Profitability will come in the next couple of years or so. Hopefully, as we get more successful, we can live just a little bit more comfortably."

During his tour, Rix showed me two new conditioning tanks, shiny and domed, with oval-shaped hatches that gave them a space-age feel, circa 1965. They're used to age beer, and were the linchpin in the plan to make that 50 percent production leap. Previously, Bent Stick had conditioned beer simply by leaving it in the fermenters, which kept those vessels from being used to actually make more beer, and therefore capped growth. Those shiny new tanks were essential to the comfortable living Kendall aspired toward and to transforming the company from its status as a relative newbie of Edmonton craft brewing into one its elder statesmen. One day, anyway. Rix pointed out that only one of the two new conditioning tanks had been plumbed into the system.

"Once we save up some money," he said, "we'll get the other one connected."

After leaving the brewery with a growler of wit for the evening, my well-rinsed bottle filled right from the tap, I drove along Fort Road toward the main drag. As I travelled, I looked over at the construction of a new transit facility, the one large project that was advancing in the area, being built on the site of the old meat-packing plant. When construction of a $210-million garage began in 2016, the relief was palpable. Scheduled for completion in 2019, it would bring seven hundred employees to the area every day. "It's hoped the new facility will bring some life to the neighbourhood," wrote one reporter after the groundbreaking.

Is progress ever easy? I thought as I headed home, the traffic having lightened slightly. No matter what stage you're at, though, there's something to be said for having what seemed to be the hardest part out of the way, which was simply getting started. If the gold rush was underway, Bent Stick at least had its shovel in the ground.

FINE AND DANDY

Three hours south, down the QEII Highway, the busy, four-lane spine that runs along the trunk of the province, Calgary's Dandy Brewing Company at first followed a path parallel to that of Bent Stick, at least in terms of modest beginnings. It did, however, pivot in a way that its Edmonton contemporary resolved not to. Now, Dandy's current incarnation is about as removed from its scrappy beginnings as a mass-produced lager is from, say, a brut IPA, a trendy style that connoisseurs liken to champagne (at least two varieties of which Dandy has released over the course of its short history). Dandy is a standout among craft breweries of Calgary's south. But that status came with a cost that, at the time I spoke with owners and staff, they were still paying.

The operation is part sixty-five-seat restaurant and part brewery, but to call it a brewpub would be too casual a descriptor. To compare its menu to common taproom fare—greasy, over-salted pretzels are, for some reason, ubiquitous—would be insulting. The kitchen, at the time of writing run by chef Merritt Gordon (who came with a pedigree having contributed to upscale contemporary menus at Montreal's Au Pied de Cochon, Calgary's Rouge, and more), doesn't even have a deep fryer. Things get surprising and fancy at Dandy. Snacks include chicken liver toast and oysters; dinners may involve ceviche and lamb's neck. Whatever you might have during a visit, I recommend starting with those Alberta trout rollmops. I'm still thinking of them.

"Dandy... truly offers some of the best taproom cuisine I've enjoyed in recent months in Western Canada," wrote Calgary-based food critic Dan Clapson in an August 2018 review for *The Globe and Mail.* "When it comes to the food—which arrives after you've begun sipping on interesting beers, such as their refreshingly puckering plum sour or the tropical-tasting wild sour ale—you can tell almost immediately that chef Merritt Gordon has experience well beyond that of the typical pub chef."

A former motorcycle garage, the "tasting room," as Dandy calls it, is a chic, modern space decorated with local art. It's located in Ramsay, one of Calgary's oldest neighbourhoods, the difference from Bent Stick's locale being that Ramsay is actually seeing results in its revitalization efforts. It helps that it's within walking distance of downtown, perhaps giving Dandy a leg up in the effort to make good on a business plan and give them an uncommon air of sophistication as they do so.

Of course, it was not always this way. In fact, Dandy's origins might be even humbler than those of Bent Stick.

IN 2016, I visited "Dandy 1," the location where the brewery released its first beer in August 2014 and established itself as one of Alberta's first bona fide "nano-breweries," too small to even qualify as a micro-brewery. The year previous at the old location, co-owner and sales director and marketing head Matt Gaetz told me they'd produced a grand total of 242.5 hectolitres. And they'd done it in what he now remembers as a "janky" brewhouse.

It's hard to fault Gaetz for being none too nostal-
gic about the original Dandy. The brewery got its start,
like many of Alberta's breweries, in a business park, in
this case in northeast Calgary, not far from the international airport. It was tucked in among a collection
of auto service shops, parts suppliers, niche consulting firms, engineering outfits, construction offices,
real estate, all of them housed in dull, industrial strip
malls. Dandy had a cozy but cramped taproom that
could seat twenty-four, and served charcuterie boards
imported from Calgary restaurants. It was a cool place
in a conservative district unused to after-hours activity.

"We're in such a weird spot," Gaetz told me back
then. "We've had police and the fire department come
by, especially when it gets dark early and the lights are
on, people are drinking beer. The police have been like,
'What the hell's going on here?'"

Still, it did the job of launching the company. Dandy
is the product of four young minds, the oldest, Gaetz,
being in his early thirties when things got started. Ben
Leon, the instigator, is a former schoolteacher who
galvanized the group in the excitement of deregulation in late 2013. Leon had been homebrewing with a
buddy, Dylan Nosal, for years. He was also connected
to Derek Waghray, a high-school friend who'd go on to
graduate in the second class at Olds College. Married
to the cousin of Leon's wife, Gaetz was a veteran of the
retail sector, where he'd worked his way up to district
manager with Calgary Co-op liquor stores, overseeing

tens of millions of dollars in product. One night, the four of them met over beer.

"Ben came prepared with a small-business plan," said Gaetz. The idea was to start as small as possible and see what happened. "At the end of the night after a few pints we were like, 'Yes, let's start a brewery!'" In what might be the equivalent of smashing open piggy banks and counting the contents, each of the future partners gave a number they felt they could contribute. "And that was our budget," said Gaetz. It came in under $100,000. Hoping to beat other startup breweries to the suddenly open market, they incorporated on January 26, 2014, about two months after deregulation.

In the early days, the brewery had a homebrewer-gone-wild feeling to it (which is presumably what Gaetz meant by "janky"). Everything was manual. The taproom was a fraction of the square footage of their entire bay, which stretched back long and narrow to accommodate a cramped brewhouse that spoke to making the most of what's at hand. Having no auger to transfer grain to the mash tun, the team would mill grain in a loft that sat overtop the taproom and then haul it down to the tanks by hand. Matt Mercer-Slingsby, Dandy's retail manager and the company's first full-time employee, remembers the operation and equipment as being, endearingly, "punk as fuck."

But it was within budget. The hot liquor tank, used to heat water for brewing, and the mash tun, where grain is steeped and its starch is naturally converted to sugar for fermentation, were born of the same vessel. Staff at another brewery put them in touch with one of the veterans of Alberta's brewing scene, Al Yule, who ran an early Calgary brewery called Brew Brothers. Yule, as Gaetz remembers, had "acres of brewing equipment," including a large kettle that Dandy purchased and cut in half to form their hot liquor tank and mash tun. For a heat source, they'd slide a 95,000-BTU burner under the hot liquor tank. As for fermenters, they sourced 350-litre plastic tanks, which had AGLC inspectors scratching their heads. They let Alberta Health Services sign off on them. The tiny tanks kept batches small and, given the quick turnover, creativity high. They also kept risk low. It didn't happen often, but Dandy could dump a batch if it didn't turn out how they wanted, said Gaetz, and it wouldn't be a great loss.

Eventually, those little tanks would prove restrictive. "There wasn't a time when a fermentation tank was left empty for any more than twelve hours," said Gaetz. "It was just one after another."

Demand outstripped capacity. They would send out a pallet of Dandy in the Underworld Sweet Oyster Stout, one of their first beers and their best-seller to this day, hoping it would last three months. It would stretch maybe three weeks, leaving Mercer-Slingsby apologizing to clients at shops and restaurants. By 2017, the team knew it was time to grow. They hoped to do so in place.

"We wanted to do the Alley Kat model and take over as many bays as we could to help us grow," said Gaetz. They got quotes on construction costs for expansion and took them to the landlords. When they balked, Dandy walked. "We wanted to make sure that we had a space where we could do our best, and it wasn't the northeast space."

By April 2018 they were serving beer brewed at the old location out of their new taproom in Ramsay while the final touches were being put on the brewery and were positioned to start cranking out four pallets of stout for every one they eked out in the old days. There was, however, an inevitable drop off in production during the move.

After transitioning to the new location, Dandy got a sense of just how tight the market for Alberta craft beer had become. When they incorporated in January of 2014, they were one of less than half a dozen newcomers to the industry across the province. In 2018, the year they'd started over in Ramsay, things had changed. Radically.

When Dandy 2 opened, it did so alongside roughly a dozen new breweries that had opened that year in Calgary alone. What they'd worried about from the start, and what motivated them to get brewing as quickly as possible, had finally come to pass: They had competition—competition they weren't quite ready for.

"When we moved breweries there was a four- to five-month period where we had little to no product at all," said Gaetz. There was enough for the new tap-room but retail sales were reduced to virtually nothing for about three months. "When we got back to brewing what we wanted to, we were kind of reintroducing ourselves to the market. And it was a completely different story selling ourselves into the market again and getting our market share back than it was in 2014. We had to fight tooth and nail to get our spots back and convince them we were back for good."

That meant getting back onto taps at restaurants and bars as well as onto crowded shelves at liquor stores. No longer was Dandy merely up against the macros; there was now a small army of Davids who were starting to watch their own backs during the battle against Goliath. If Dandy wasn't able to keep a steady supply, "there are six guys lined up behind you who can," said Mercer-Slingsby. During the shortage, the salesperson spent his time making courtesy calls in the hope that courtesy would later be reciprocated by buyers who'd had to look into alternatives.

"Even if I don't have product, I show up," said Mercer-Slingsby. "I say, 'Hey man, we're still here. Please don't forget about me.'"

As a result of the disruption, Dandy was still trying to find a balance between production and sales in early 2018, said Gaetz. That January, they'd made 240 hectolitres and sold only 130; in February, they narrowed the gap, making 130 and selling 111. It was beer math that illustrated business realities that had changed a lot in a short period of time.

"We have to sell *x* amount of beer to make this actually work, and sometimes it's not there and sometimes it is," said Gaetz. "The stressful part is making sure that it is more than it's not."

The new location and its challenges galvanized a sense of business maturity among the four partners. In Edmonton on the night of the annual Alberta Beer Awards in March 2019, I caught up with Gaetz and Mercer-Slingsby at Craft, a cavernous restaurant in a quiet section of the downtown business district. We each sipped tulip-shaped glasses of the sharply fruity and vaguely white wine–like Dandy Wild Sour. The pair looked confident and sophisticated, dressed in patterned shirts buttoned to the collars, ties tightened, sleeves rolled up just above the wrist. Mercer-Slingsby wore suspenders. I noticed that Gaetz's hair had a touch of grey in it that I didn't remember from when I'd seen him a couple of years earlier.

They'd put no beers forward for consideration in the awards but were attending anyway, more eager to be present than to be recognized. Mostly, they just wanted to step back from the brewery business for a few hours and celebrate it instead. It was a rare moment to relax and take stock. When their beers first appeared on shelves, notably the oyster stout and a golden-brown ale, I recall feeling that the bottles were an announcement that change was coming for Alberta craft beer. It was obvious in the whimsy of the art on the labels and the flavours of the beers, which were as bold as the shirts on Gaetz and Mercer-Slingsby's backs. And now, the young upstarts were redefining the experience again with a tasting room for a taproom, a tacit challenge to the province's craft beer community to already reconsider what it had become and where it was going. The definers were already redefining.

They said none of that, of course. They may not even have agreed with it. What Gaetz did say was, "You didn't expect [this] from day one—you know, a bunch of goofballs making beer, seeing what we can do with it." Like Leon, he'd walked away from a job with a pension and benefits, and at a time when his wife was expecting their first of two daughters. "The biggest thing is taking that leap," he said.

At the time of writing, there was no conventionally happy ending to Dandy's story. While there's nothing to suggest that it won't, the gamble has yet to pay off. "We're in debt," Gaetz said with a slightly grim laugh, "and probably will be for the rest of our career."

That, too, is shaping Dandy as it adapts to the realities of the rapidly expanding market. The direction is remarkably pragmatic for a gang of "goofballs."

"We still want to grow," said Gaetz. "We want to make sure that we're relevant in a very busy province and do what we do—and what we're good at doing is making beers." As veterans of Alberta's modern era of craft beer, Gaetz feels Dandy has largely proven itself among fans of local craft beer. "Reaching a different demographic, a different market, that's the challenge right now."

The next step, added Mercer-Slingsby, "is putting yourself in a position where you can be all things to all people."

Most seasoned entrepreneurs would shudder at those words. Small-business experts advise zeroing in on your core customers and being as loyal to them as you want them to be to you. Not even multinational mass-producers try to broaden the horizons of resolute, lager-drinking customers by suggesting they give IPAs a chance. Instead, they tend to make forays into the craft market through acquisition, and therefore almost clandestinely, buying up small-batch brewers who've developed solid brands (such as Molson's purchase of Ontario's Creemore Springs in 2005; Creemore Springs, in turn, bought Vancouver's Granville Island Brewing in 2009; in spring 2019, Wild Rose Brewery was purchased by Ontario's Sleeman Breweries, which is owned by Sapporo). Even Big Rock backtracked on an attempt to go head to head with the big brewers in the lager category.

What might make Dandy different in its efforts at mass appeal is that it has never given up on trying to satisfy the growing thirst for weird beers in Alberta. This is a brewery, after all, that once collaborated with BC brewery Twin Sails to produce a sour guava milkshake IPA, a tongue-in-cheek attempt to roll up every beer trend of the day. But Gaetz and Slingsby know that, while customers may be loyal to craft, they tend not to ally themselves solely to a single brewery. There is no Dandy Man in the same way there might be a Bud Man. A surfeit of choice has led to a craft beer culture that encourages consumer promiscuity, a relentless search for satisfaction that will follow new beers almost regardless of who makes them. With Alberta's craft breweries competing for what's estimated to be just 10 percent of the province's beer-drinking market, that fickle nature could sink a too-idealistic business plan. Those customers may go, but they're almost certain to come back. In the meantime, target another.

"To grow the way that we want to grow, we need to have a little bit more of an approachable style, maybe more of an approachable beer," said Gaetz.

The answer was a lager, and one not too crafty. Rather than go the route of other craft brewers who've added a Euro-style, hoppy lager to their offerings, Dandy kept it simple. "It's got a lot of toasted rice," said Gaetz. "We went more of the Japanese-style, Sapporo route." Added to the core lineup, it's clean, crisp, and crystal clear. "We put a lot of care and work into the lager."

Which is to say they've put a lot of care and work into their future. Gaetz joked about how the new addition affected his dad's perspective on the business.

"'Matt's got a brewery. That's fun,'" he said, as if mimicking the elder Gaetz. "But now that we've got a lager, my dad's like, 'Oh, Matt's making *beer* now.'"

The truth of the matter is that Alberta craft beer may be missing out in not targeting Gaetz's dad and other boomers. The generation has some of the highest spending power and most leisure time in North America. It's a market a brewer ignores at its peril, particularly one such as Dandy, which was hoping to get packaged product into BC by summer 2019, following on recent success with placements in Saskatchewan.

All the talk of lager while we drain the last of our sours at Craft raised a question for me. Did Gaetz think the reasons he and his partners got into this business are the same as the ones that keep him in it? The subtext: You started with a passion for brewing unique beers and now business is driving you to make a toasted-rice lager. He paused briefly, then said, "I think they are different." Then he stopped and started over. "Well, no," he began.

"The fundamental thing that got us together was building toward something that we believed in—as corny as that sounds. But we were four guys who didn't know each other all that well and we all had this vision of building this craft beer community around us. I don't think that has changed. That's what keeps us going. That still holds true today. It's just the way that we go about doing it has changed."

Gaetz paused to reflect before speaking again. For

him, the question was a lot to consider. What do you do? is easy to answer. Why? is not. The answer to that is tied close to identity, which is anything but objective, and therefore hard to land on. Gaetz narrowed his eyes and gave me an imploring look.

"Does that make sense?" he asked.

LOCAL MOTION

The post-2013 brewing scene has been animated by many things, such as the energy and drive of younger entrepreneurs like those at Dandy and Bent Stick. But another thing these brewers are demonstrating and using to their advantage is a love of all things local. In a way, this phenomenon, particularly among such a considerable swath of beer consumers, is nearly miraculous. That's because of another act of disruptive deregulation that happened more than a quarter-century ago.

Beer lovers of a certain age in Alberta, and those of wine and spirits, likely remember 1993 as a very good year. That September, the Government of Alberta, led by the late Premier Ralph Klein, announced that it was exiting the liquor retailing business. The motivation was money. The government would save $67 million a year if it divested itself of its 208 stores, which employed 1,500 people. They could find jobs in the privately run shops that would take over, Steve

West, minister responsible for the Alberta Liquor Control Board at the time, assured Albertans. Some of those employees would even be given the option to buy the shops. Six months later, the province's last government-run liquor store shut its doors forever.

Today, those government stores have been replaced with roughly fifteen hundred privately owned stores. More important to note, though, is the difference in what those stores now contain. Before privatization, 2,200 products were available to Alberta consumers. There are now more than 26,000.

For Alberta brewers today, that statistic is good and bad. To the bad, their products compete with the best that retailers can get their hands on, from all over the world. A locally made pilsner, for instance, takes on the famous and delicious Czech lager, Pilsner Urquell, which dates to 1842. Beer critic Michael Jackson considered it one of the world's best representatives of the style.

Privatization also introduced another complicating factor for the newly established Alberta brewer: sheer quantity. If a brewer, for example, made its debut with a single beer packaged for retail sale, that beer becomes but a drop in a sea of the stuff. Currently, there are approximately five thousand beer products available to Albertans. Should that new brewer not have sufficient marketing power to give that beer a voice, which is likely the case, it sits on liquor store shelves waiting for the consumer to grow tired of predictably excellent European lagers or, for that matter, cheaper but familiar ones. No store stocks every beer product, of course, but that single Alberta-made beer nevertheless stands a one-in-dozens chance of being picked. Those odds would have been much more in a local brewer's favour before September 1993 (though there weren't many local brewers).

Seeing the good side of privatization requires only a slight shift of the prism to shed a different light on the nature of the consumer. The rapid ramping up of selection that accompanied privatization had an educational effect, in that it enabled the rise of the beer geek. In that way, it prepared the province for the deregulation that would follow twenty years later, and which permitted the proliferation of domestic manufacturers we see today.

"When the rules changed in 2013 it was a natural evolution that kind of opened the floodgates," said Edmonton-based beer writer Jason van Rassel. "Even though we only had a dozen or so breweries, what existed prior to that was a very sophisticated marketplace. We had some of the best beers from around the world available to us on the shelves of our retail stores in Alberta because of privatization."

In other words, the privatization of 1993 helped Albertans train their beer palates so as to be ready for the deregulation of 2013. A visit to the walk-in fridge of a decent liquor store during that twenty-year period was an educational experience. It alerted consumers to the fact that shopping for beer wasn't just a matter of picking between macro-brewed lager A, B, or C. It taught us that beer came in myriad flavours and colours. Which meant we were also open to being taught, eventually, that there were alternatives made by local producers.

Being a free market, however, meant the privatized system as a whole did not have the same incentive to champion homegrown product as would a system run by a government seeking to develop local supply chains or simply secure existing jobs. Instead, cheerleading became the responsibility of the community, a community that has had more impact on the growth of Alberta craft beer than any government decree. Policy may change rules, after all, but rarely does it have the power to change culture.

THE HUNGER FOR LOCAL

That same sense of community has branches that reach out far beyond craft beer, though they are all connected to the same cultural tree. They support each other as a result. The arrival of local craft beer in Alberta could not have been better timed, given the broader celebration of locally produced food. This has been a long time in the making, and it's meant breaking with what once was ingrained as a better way of living.

Convenience food—processed and packaged for quick and easy consumption—long ago began to compete with whole foods as a staple in North American diets. It was almost no contest. Countless factors conflated to make fresh seem an outdated concept, the stuff of less enlightened times. Prepared meals were an innovation of the Second World War when food had to be fit for the battlefield. Thereafter, increased urbanization put more people farther from farms. More families had both parents in the workforce, limiting time and sapping energy, while also increasing income to spend more on effort-saving suppertime solutions. Microwaves began to see mass-market distribution in the decade following the Second World War. In 1954, the US company Swanson launched a line of TV dinners in compartmentalized tinfoil trays and sold twenty-five million of them in the same year.

Karen Anderson, owner of Calgary-based Alberta Food Tours and co-author of *Food Artisans of Alberta: Your Trail Guide to the Best of Our Locally Crafted Fare*, believes that decades of embracing heat-and-eat food culture has come at a cost. After twenty-one years as a nurse practitioner, Anderson decided to refocus her efforts to improve people's health through prevention.

"I found that the further people got from food, the sicker we were getting as a society," she said. "I thought, 'How could I affect them at the front end? How could I change their connection with food?'" She started her company in 2006 and began to link consumers to food sources by showcasing the farmers and producers who make stuff that won't be found on shelves in the frozen food section. "It was a chance to influence people's choices."

Over the years, Anderson has taken groups to visit local-focused restaurants and businesses in Calgary neighbourhoods including Kensington and Inglewood, and organized festivals that celebrate farm-to-fork cuisine. She feels that she's making an impact by tapping into the fact that people are beginning to question how much to trust their diets to corporations beholden to shareholders.

"I think people's values are changing and people are realizing that how they spend their dollar determines the food system we get," said Anderson.

In 2016, Albertans spent nearly a billion dollars on food at farmers' markets, making local products more than a niche item targeted at high-income earners. Consumer-supported agriculture is also proving to be fertile ground for agri-preneurs. In 2018, CBC.ca reported that fifteen Alberta farms offered the fresh-food-by-subscription service, compared to just three in 2008. Anderson may be one of a relatively small number of provincial advocates for the trinity of, as she puts it, "strong soil, strong food, strong people." Nevertheless, she's tapping into a global movement to counter the globalization of our food, a movement that has been documented in best-selling books such as *The 100-Mile Diet: A Year of Local Eating*, by Alisa Smith and J.B. MacKinnon, Jennifer Cockrall-King's *Food and the City: Urban Agriculture and the New Food Revolution*, and the work of food journalist Michael Pollan. All of these focus not only on what's lost with decreased access to local food but on what's gained by finding ways, as Pollan has put it, to "eat the view," or maintain idyllic spaces by ensuring they're too important to food production to give over to commercial or residential development. A similar sentiment is expressed in the menus of top independent restaurants. They frequently lean hard on what's become must-have marketing: locally sourced ingredients. I asked Anderson what all of this might mean to local craft beer.

"Along with that comes the realization that Alberta grows some of the world's best barley," she said. "It goes along with the same values that people would rather support an Alberta enterprise than something that comes from a factory in Toronto or Milwaukee." But, she added, "I don't think that people think, 'If I buy this local beer, I support this local farmer.' I don't think it's the same motivation. It might be more taste driven."

Likely, Anderson's right. To a point, at least. While awareness has grown about the quality of Alberta barley, the local farmer still toils in obscurity. No one, for instance, shows up with sacks of organic malt for homebrewers to snatch up at the farmers' market. It's also difficult to apply Anderson's lens of better health to better beer, given the links between alcohol and addiction, liver disease, some forms of cancer, and a host of other ailments when consumed in excess.

But during each of her food tours, and throughout her guide to Alberta food, she tacks on "local watering holes," leading people toward better ways to imbibe. They're a good fit. Even if Anderson feels consumers' motivations for choosing local beer are not yet the same as those for choosing local food, it's easy to see how attitudes could shift. Barley farmers may not be setting up a booth at the weekend farmers' market. But the brewers are. And they're there just as much to tell their stories as they are to sell their beer, putting a face to a product in a way that just isn't possible with imports.

If consumers aren't quite ready to understand the impact of their beer-buying habits on local brewers and farmers, some members of the food community are eager to show them. Locally sourced ingredients have become so much an expectation of customers that it may be at risk of losing its impact as a differentiator. Stocking a good selection of Alberta craft beer, however, may still have the power to make a restaurant stand out.

Darren McGeown took that to an extreme with Arcadia, a craft beer bar west of downtown Edmonton on 124 Street, a low-key alternative to the raucous entertainment strip that is Whyte Avenue. Named after an ancient Greek depiction of a utopia, the place may strike fans of Alberta beer as just that, as the taps and cooler feature beer from nowhere else. The food, however, by McGeown's own admission, may not encourage the same reaction. You wouldn't quite call it craft.

Since opening Arcadia at the end of 2014, in the earliest days of the Alberta craft beer boom, McGeown has more or less run the place by himself. When we spoke, he was in his mid-thirties but wizened by what he called the "Irish curse" of early grey hair and the stresses of entrepreneurship, which in his case includes getting the groceries, cleaning, marketing, and even doing the cooking. He did not reference any emphasis on locally sourced ingredients.

"We do food but it's definitely not our focus," said McGeown on the quiet night I visited Arcadia. A couple of guys watched a hockey game on a TV above the bar; another pair chatted at a table across from us. On other nights, the owner assured me, there may be music, variety shows, burlesque, comedy, and cask nights that have the place hopping. It keeps McGeown stuck in the tiny kitchen in the back. "At the end of the day, it's hot dogs and pizza," he said with a shrug. "Pretty much what I'd eat at home."

Instead, McGeown keeps the focus on Arcadia's "atmosphere," which is largely the result of the beer. The bar has a reputation for being like a box of chocolates when it comes to local lager and ale, in that you really never know what you're going to get. Instead of the safe bets of Blindman, Apex Predator, and Wild

Rose that tend to dominate beer lists in most other restaurants, McGeown is a curator who doesn't have the heart to leave people out. Not that he doesn't work at landing the latest and greatest; he listens to customers and industry colleagues and connections about kegs he should try to get his hands on. But McGeown will also hear out a brewery owner who pops by looking for someone to take a chance on his or her brew.

"If the owner of a brewery comes in here, I'm like, 'I gotta give this guy's beer a shot, because he's going through the same battle I'm going through,'" said McGeown. "Could I bring in a better beer? I can. But he's a startup business owner like I am. We've got to fight the good fight together. And I want to be a reason that, hopefully, that person can be successful. Or at least get started."

Sherbrooke Liquor has taken a similar approach since it discovered that craft beer was an unfilled niche in the city. As David Owens told me, his brother Jim Pettinger set the focus not long after the store opened in a north Edmonton strip mall more than fifteen years ago. "He's always been a bit of a collector in his life and he decided to start collecting beer, basically."

The store's walk-in beer fridge came to contain hundreds of beers by the time Pettinger stepped away as store manager in 2012, and the store's name was made as *the* place for beer lovers. I joked to Owens, owner and chief operating officer, that I won't visit without my kids; their getting cold and bored is the only thing that gets me to the till and out the door in a reasonable amount of time. He laughed in a way that suggested my strategy wasn't unique.

These days, the beer fridge is managed by Stephen Bezan, whose work has earned him an Alberta Beer Award for community building. Those efforts have also played a role in Sherbrooke being named best retailer at those awards in 2018 and 2019, and being listed among ratebeer.com's "world's best places for beer" in 2016. When I talked to him and Owens in spring 2019, Bezan said the fridge contained roughly nineteen hundred different beers and was on track to hit two thousand by the end of the year. "We're constantly trying to use every square inch possible," he said. At twelve hundred square feet, it's now a tight labyrinth of bottles and cans, with the shelves stocked by beer style from floor to ceiling.

The only sets of shelves not organized by style are the three or four directly inside the entrance. Bezan reserves these for Alberta breweries. In 2015, he said, the section contained a little more than 90 different beers. Four years later, they were stuffed with 618.

While that kind of industry support alone could have qualified Bezan for that award for community building, the real boon to breweries has been Sherbrooke's growler bar. The six taps change weekly if not daily, and sales have dictated that it be dedicated almost entirely to Alberta brewers, whether they are established or just entering the market.

"I see it as a way of showcasing a brewery," said Bezan. "Almost being like a launching point."

It was there that I discovered a brewery that I'd learn was just minutes from my home. SYC (pronounced *sick*) Brewing opened in January 2019, in a light industrial area in west Edmonton. That's one of the reasons co-owner Richard Fyk didn't base the business plan on its spacious and comfortable two-level taproom. Given the location, which is further isolated by being hived off from most of the city by broad and busy 170 Street, no one's going to find SYC by accident—unless they happen to find it on the growler bar of a centrally located liquor store.

Fyk told me that for a new brewery like SYC, Sherbrooke acts as a kind of resumé builder. Apart from a place like Arcadia, most restaurants and bars are hesitant to take a chance on a newbie brewery. "They don't want to take that first step," said Fyk. "They want to see you out there." Sherbrooke's growler bar, then, is a foot, or maybe a toe, in whatever door is willing to open a crack.

Out-of-towners have also benefited from the Sherbrooke bump. For Patrick Schnarr, owner and brewer at Calgary-based Outcast Brewing, the store was his entry point to the Edmonton market when he launched his first beer in November 2016. After Bezan put it on tap (Sherbrooke was the first business to buy an Outcast keg, said Schnarr) the brewery became a mainstay on the bar, sometimes selling out in a single day. That support played a part in helping Schnarr make the leap from brewing on other breweries' equipment to opening his own place, in the latter half of 2019. In the lead-up to that he estimates he moved fifty kegs through the shop and two hundred cases of cans. Every batch has been unique; Outcast has no core beer, making the brewer, perhaps appropriately given the name, a respected oddity in the Alberta market.

"Stephen has pushed our products as some of his favourites and we have developed an awesome following in Edmonton as a result," Schnarr told me by email. "Sherbrooke has been instrumental in our success [and] our number one account since the day we launched."

"That's something that I'm passionate about personally, to help out," said Bezan. "It's tough to get your name out there. I'm very happy to be [a brewery's] first foray into the market."

Alberta beer still needs champions like Bezan and Sherbrooke. As the sparse patronage at Arcadia the night I visited might suggest, it's not yet inciting riots on its own. "Obviously I'd like to be busier," McGeown told me.

The first time I interviewed McGeown, in 2016 for a blog post, his stance on going exclusively local was highly principled. "Right now, we're coming through tough times and we need to be supporting each other any way we can," he said. "There's no better time than now to be expanding into other industries and putting more money back into Alberta. [Carrying] this beer, I'm helping to create jobs for people in these breweries and new brewpubs. If one brewery succeeds, it means another one opens, and that creates more jobs."

Three years later, those principles were still clear. In other ways, McGeown had backed down as a business-person leading with his values. Arcadia had opened not only as a bar devoted to Alberta beer, but one with a vegan menu. He's since brought in meat, a change encouraged by customer requests that coincided with the opening of nearby independent deli called Meuwly's that features house-made meats and pre-serves. He even dropped the dietary restriction himself. On the beer, though, McGeown won't budge. Should a customer walk out of his place because he doesn't have a macro-beer on hand to suit their tastes, so be it. In that case, McGeown is guided by something big-ger. According to that Greek myth, Arcadia is a perfect but lost society, impossible to reproduce. If McGeown acknowledges that latter point, he doesn't let it sway him. Arcadia, both the restaurant and the place, still represents the kind of community he wants to pursue. For him, beer is the way and for now business is good enough. "I'm happy where I'm at," McGeown told me. "There's helping the scene grow in Edmonton and Alberta, and that's what I would like to do. There are other riches than money."

AS MUCH AS McGeown's embracing of local beer is a move toward an ideal state, the current enthusiasm for things made nearby may also represent a desire to move *away* from other things. Social disconnection may be one of them. McGeown grew up in a smaller city near Edmonton, where people knew the name of the baker or the guy who ran the sports shop. "It's like that in Ireland," said McGeown, whose parents were born there. "Everyone knows the butcher. The butcher's like a celebrity. There's a great sense of community."

That craving for community may be stronger than ever. Clear links have arisen between high social media use and feelings of loneliness, the antidote being interaction with real live human beings instead of their accounts and avatars. Isolationist, nationalistic ideologies may also have us pining for togetherness. Since Donald Trump became president of the United States in 2017, the country has led with an "America-first" mentality that has seen it withdraw from all manner of international partnerships, including the Paris Agreement to combat climate change, the free-trade pact of the Trans-Pacific Partnership, and NAFTA, to name but a few. In early 2018, a poll found global approval of US leadership to be just 30 percent. In Canada, another poll at the time put that number at 13 percent (except in Alberta, where 29 percent approved of Trump's leadership). Such cellar-dwelling statistics may not be direct admonishment of one man's attempt to be an island while dragging more than three hundred million people along with him, but they might be taken to indicate that we struggle to see the sense in refusing to be a part of a community, large or small.

When I spoke to Rieley Kay, co-owner with his wife Kim of Cilantro and Chive in Lacombe, I didn't subject him to these pet theories about the human condition. I tried another one on him instead. What if the increasing love of local as we know it today, I said, is tied to the recent recession? Is it possible that, despite the premium we might pay for things made or grown in small batches in Alberta, we might be willing to pay for them because we are also increasingly coming to recognize that *we need each other*?

Kay kindly considered the idea, but wasn't sure the recession should be credited with being so significant a motivator. His experience suggested Albertans aren't quite so reactionary. They're better than that. The idea about community—that one might actually apply (though Trump probably isn't behind the Alberta craft beer boom).

After a few years of catering, Kay opened a restaurant in Ponoka in 2012. He wanted to feature the output of local farmers and craft beer was a natural extension of that. Even if not much of it was sourced from Alberta at the time, the barley it was made from was. When their lease in Ponoka ended in 2015 the Kays relocated to Lacombe, a largely agricultural city of approximately thirteen thousand people. They opened Cilantro and Chive right downtown, across the street from the offices of the Field Crop Development Centre, where new barley strains are developed and tested. The menu shifts with the seasons and what's coming off the farms, be it bison, elk, really almost anything.

"It's a fun challenge with some of the local farmers to highlight what they have and get consistent products," said Kay, who's trained as a Red Seal chef. When I asked him if he was speaking euphemistically, he laughed and insisted that "it's really fun to see our staff get together and talk items through and collaborate on different platings and different items."

The new space, combined with the current surfeit of local craft choices, also allowed Kay to expand the restaurant's beer list to the 165 varieties it offered at the time of writing. It's not exclusively Albertan but he guessed that provincial brewers accounted for about 80 percent of it. "Our big push is on the Alberta breweries for sure," said Kay, who took Coors Light off the menu despite it once being his top seller. In any case, that percentage was enough for Cilantro and Chive to factor into its wins for best pub/restaurant awards in both 2018 and 2019 at the Alberta Beer Awards, despite big-city competition.

Kay did believe, though, that if there was any connection between the downturn and growing interest in all things local, it might be in how people are spending their free time. That is, they may be spending a little less money on it. Lacombe is located just an hour and twenty minutes from Edmonton and an hour and forty from Calgary. Perhaps decreased disposable income has reshaped the average Albertan's sense of tourism, Kay suggested. "We have seen a lot more people doing day trips," he said of the restaurant's clientele. "People are more focused on what's in their backyard and they're taking time to explore that and see it in a new light."

In his view, neighbourliness is independent of economics and the health of the energy sector. "We want to contribute to our community," said Kay, "we want to contribute to the people who support us. They're our neighbours, they're our friends. For us, it's important for a community to grow together."

What is a community if not a group of people who feel accountable to one another and responsible, on some level, for the collective sense of well-being? It comes back to McGeown's celebrity Irish butcher, the link between consumers and producers who may never meet but want to feel somehow connected to each other. It's like a redefining of comfort food. "A lot of people are concerned about where their food's coming from," said Kay. "They want to get to know their suppliers." He saw his menu, and the beer list, as part of the connection.

As Karen Anderson told me, the kind of society we want to be is in part determined by what we eat and drink. She believes that people, Albertans among them, are coming around to this idea. "It's just common sense," she said. "But it results in a rich, rewarding culture for us." Craft beer is therefore two things at once—a really tasty way to quench your thirst and a symbol of community cohesion.

WILL THIS INTEREST in and support for local last? Are we on our way to realizing that rich, rewarding culture? In recent years, Alberta craft beer reached two important milestones that suggest we're making those strides. In 2017 it finally made a breakthrough: It was served at the Calgary Stampede. Behind the coup was the Alberta Small Brewers Association, then headed by executive director Terry Rock.

"It started with a social media outcry the year prior," said Rock. Back then organizers of the ten-day rodeo and festival were bound to an agreement with Anheuser-Busch InBev, makers of Budweiser. When the contract came up for renewal, Rock acted on behalf of his members.

"I did a lot of media," he said, "and just used the opportunity to talk about how important local beer is and the fact that consumers want local beer wherever they can get it." After a few months of conversation with the Stampede, "we came up with a plan." Beer from twenty-three Alberta small brewers—including veterans such as Big Rock and Wild Rose, but also post-deregulation breweries such as Calgary brewpub and distillery Last Best and Cold Garden Beverage Co.—

ended up being offered alongside Labatt products in the Big Four Station, a venue named in honour of the four men who founded the event in 1912, one of them being A.E. Cross, original owner of Calgary Brewing and Malting Company. After that, said Rock, "Northlands became bold enough to try to go exclusively local beer." Between July 19 and 28, 2018, the non-profit organization's annual K-Days festival only sold beer from twenty-six Alberta craft breweries (and eleven local spirits).

To him, both organizations' commitment to local showed "how this is a market-driven phenomenon. Consumers are demanding better product and they're identifying with it—at an emotional level."

Craft made a comeback at both events in summer 2019, but the future made no promises. "These agreements are renegotiated every year so we don't have any guarantee of these partnerships continuing," said Lauren Reid, Alberta Small Brewers Association marketing and events manager. "But we are hopeful to keep these relationships going, as we see a lot of value for both sides."

The ability to see that kind of value, and so to sustain the enthusiasm for local products in general and for craft beer in particular, is going to come down to the most basic factors. The most important is quality. Just as the privatized liquor market blessed Alberta with the burden of choice, it raised the expectations and sophistication of beer drinkers. Greg Zeschuk, founder of Blind Enthusiasm, which includes two breweries in Edmonton, sees that as a positive.

"When we started [making] unique beers, it wasn't a big stretch for people to enjoy them," he told me. "So, there was the opportunity to talk to [a market] that had appreciation for different beers and then start seeding our own into it."

He believes that consumer attitudes are such that "if you can create the same quality locally, you'll always succeed." There's a warning built into that statement. Showcasing craft beer at mainstream events for largely unindoctrinated attendees comes with a risk.

"Ultimately, local beer wins—as long as it's good," said Zeschuk. "Just because you're local doesn't mean anything. You have to make good beer."

THE GEEKS SHALL INHERIT THE WORT

Anthony Bourdain despised beer geeks. He saw them all as the hipsters he once observed dithering over drinks in a San Francisco bar. "The entire place was filled with people sitting there with five small glasses in front of them, filled with different beers, taking notes." Bourdain told thrillist.com in 2016, "This is wrong. This is not what a bar is about ... It's not to sit there fucking analyzing beer."

Bourdain had a point. Unless it's part of a competition, beer doesn't need anyone to devote any time to its consideration once it's out in the world. Then again, Bourdain never had the chance to meet Garret Haynes, head brewer at Troubled Monk and a member of the first graduating class from Olds College. Sure, privatization in 1993 helped open up our palates, deregulation in 2013 allowed Alberta to go craft, we've got the barley and the entrepreneurial young go-getters, the local movement is in full swing—this is all fantastic for craft, but Haynes might just be a class of beer geek upon which the future of the industry depends.

Talent is one reason for revisiting Haynes's story. He has played no small part in establishing Troubled Monk as one of Alberta's standout breweries and a winner of a considerable collection of regional, national, and international awards. But it's more than Haynes's skills in hitting upon perfect combinations of malt and hops that's helping to set the tone for Alberta beer through the output of Troubled Monk, which, following several expansions, now sits at about five thousand hectolitres a year.

That's because rivalling the brewhouse in importance to the success of the brewery is a tiny room found up a set of stairs behind the taproom. Haynes calls it the "Department of Science and Fermentation, a.k.a. Ministry of Muggle Magic," a nod to Harry Potter, of course (Haynes loves to read, and has found a way to install a reference to The Chronicles of Narnia in another part of the brewery). For Troubled Monk, it's the seat of quality control.

"This is something that most breweries our size are only starting to do," said Haynes.

Considering the volume of beer, the department is modest to say the least. Picture an average apartment kitchen, some ten feet long and six wide, well lit by afternoon sun and augmented by fluorescent light and packed with gadgets. Though it's staffed by a dedicated quality control technician, Haynes knows every inch of it. It is, after all, the result of his request. The lab was "a big ask," he said, as the brewery rapidly expanded. It was the only way he could alleviate the stress of being responsible for maintaining the quality of a brewery that tied for third place as the best in the province at the 2018 Alberta Beer Awards.

"The lab is what helps me sleep at night," said Haynes.

When I visited, Haynes pointed out a device that can pierce a can or bottle to determine its oxygen content, an indicator of how quickly it might go stale. He keeps a crowded beer "library" on shelves along one wall, production-line samples that date back months and that he checks periodically for freshness. Other equipment in the lab can be used to investigate the possibility of more nefarious problems. Petri dishes, plated in a fume hood beside the library, warm in an incubator the size of a microwave to reveal any microbial fauna present in the beer or the brewery. Should there be a problem, Haynes and his staff could swab their way through the facility, like sleuths in a kind of CSI: Craft Beer, to pinpoint the source and scour the scene clean. Haynes even had a machine that deals exclusively in analyzing DNA. Instead of stopping at identifying organisms, it picks up on genes they might contain that produce one undesirable substance or another. The lab isn't entirely high-tech. In place of an autoclave, an all-important device that ensures the sterility of equipment and materials, a modern pressure cooker sat on the floor. It did the job, Haynes assured me.

In light of the room's contents, it seemed appropriate that Haynes had stencilled "Muggles Magic" on the door. For one thing, doing so saved the seriousness of the work that goes on inside from a kind of corporate science gloom. But it also suggested a striving beyond expected capabilities. In J.K. Rowling's Harry Potter series, Muggles is the nickname given to humans with no magical abilities. By applying it to beer, Haynes was walking a line. On the one hand, it was an acknowledgment that the outcome of what happens in the fermentation tank may be considered by some to be somewhat mysterious. From a creative standpoint, that's almost a necessity—no artist, or magician, should be expected to explain their processes.

On the other hand, however, it was an outright refusal to accept that an explanation might not be possible. The name was the suggestion that magic in the hands of Muggles was the stuff of science. What in lesser hands was an unknown could and should be harnessed and understood. If ever there was a place for beer geek, this was it.

For Troubled Monk's reputation, and that of Alberta craft beer, that distinction is every bit as important as the decision to have entered the industry in the first place. Craft beer is art *and* science. And that alchemy is leading us to some fascinating places.

SOON ENOUGH, it would take Darren McGeown to fascinating new places, too. Days before this book went to press, the restaurateur told me his west Edmonton bar, Arcadia, would close in March—but to be reborn. He was set to open Arcadia Brewing Co. just a few blocks away.

After years of serving great Alberta beer, McGeown felt some of it should be made in a brewery of his own. But he'd work to preserve the spirit of its progenitor. Taproom food would still feature local suppliers. He'd support arts events and local fundraisers as he always had. And he'd boost Alberta beer. In addition to taps for his beers, he'd save some for, essentially, his competition. Not that he'd consider them such.

"I don't feel like I'll be competition to them," said McGeown. As one of them, he added, "I feel like I'll still be supporting them."

4

ROLL OUT
THE BARREL

*A new generation of
brewers claims its place
in Alberta beer*

THE PARTY GETS STARTED

Just as the show was about to get underway, the emcees threw the festivities briefly into neutral. The night was young but the crowd had begun to push at the boundary between boisterous and raucous. Before things went any further, the brewers needed to be taught a lesson.

It wasn't a reprimand, though—it was actual teaching. It was March 15, 2018, and a few hundred brewers, industry reps, and craft beer lovers had gathered in the old Palace Theatre, on Calgary's historic Stephen Avenue, for the first-ever Alberta Beer Awards. A tradition was being born for an industry that was coming into its own; thanks to its newness, no one really knew what to do with it. The added challenge was that a lot of beer had already been drunk. Maintaining order early seemed a better strategy than trying to restore it later.

The emcees, a Calgary magazine editor and a beer writer who'd by then sampled some nineteen thousand brews, shouted instructions into a microphone. Come up, grab your award, smile for a photo, leave the stage. To help get the message across, an impromptu troupe clowned its way through a dry run, like airline attendants demonstrating seat-belt buckling in time to a voiceover. About as many people paid attention as might pre-flight passengers. Everyone was simply too excited about being in each other's company, and about the fact that there were finally enough people in the industry to make for the kind of party where you had to shout to be heard, even if you had a microphone.

Nearly sixty breweries submitted roughly 320 beers in twenty-two style categories, including everything from a "patio" beer to farmhouse ale to that vanguard of craft beer, the India pale ale. They were evaluated by local pros, including some of the most highly certified beer judges in Canada. Five years earlier none of this would have been possible. Five years ago, what these people were doing now, making and selling small batches of beer, would actually have been illegal. Now, what once would have been outlaws were being celebrated as aficionados.

That said, no one in the room would have admitted to seeing themselves as the latter. The dress code, or absence thereof, offered proof. With few exceptions, the most painstakingly dressed men were the two oyster shuckers laying glistening molluscs onto a bed of chipped ice. Dressed completely in black as if hoping to disappear into the dim surroundings, they stood out for their fastidiousness (a good quality in an oyster shucker, I imagine). Before my friend Evan and I had left his house for the event that night, he'd asked what to wear. I'd shrugged and suggested button-up shirts and dark slacks. We overdid it. Apart from the roughly half a dozen staff members from Calgary's Cold Garden Beverage Company—who showed up in Hawaiian shirts, khaki shorts, and wide-brimmed straw hats—most attendees stuck to something that might be considered to be one step above brewhouse-casual, but only because of the lack of workboots. Hoodies and ball caps were pretty much *de rigueur*. While the event was auspicious, it refused to be pretentious.

Beer can be, and now often is, as sophisticated as wine. As beer writer Michael Jackson once argued, during an interview with NPR in 1994, there are as many flavour and aroma compounds in a good glass of lager or ale as there are in fine wine; he wasn't about to let beer be considered the bumpkin of adult beverages. "The two products are equally complex." But never is beer thought to be stuffy, inaccessible or exclusive, regardless of how it may have been elevated by the talents of a gifted brewer. For the most part, even the snobbiest of beer snobs craves a cold one, no matter how impressed she or he may be with its colour, aroma, balance, or mouth feel.

Another reason for the casual nature of the Alberta Beer Awards, and perhaps the most salient reason, were the industry's hardscrabble origins. The Alberta craft pioneers that we've already looked at, like Big Rock and Alley Kat, endured dark times as they worked to escape Big Beer's shadow, and then only to bask in a relatively small glow of public affection. Once an underdog, perhaps always an underdog.

But now those dark days were over. Here, the industry was putting itself in the spotlight, staging a gala without the glitz, a ball where every beard would have covered up the bow ties no one was wearing anyway. That said, it was hardly a navel-gazing affair. In a way, it had already been validated nationally. At the 2017 Canadian International Beer Awards, half of the medals went to Alberta brewers who'd proven themselves capable of producing exceptional representations of beer styles derived from brewing traditions from around the world while doing it with a Prairie flair.

The awards given out that night at the Palace highlighted the same ambition. Calgary industry veteran Wild Rose Brewery won the top prize for European-style pilsner. The rustic but flavourful farmhouse ale category, which points to traditions of Belgium and France, went to a Lloydminster outfit called 4th Meridian Brewing. Edmonton's Elbeck Brews took the extra-strong beer prize for an imperial stout, a formidable style once favoured by the Russian court of Peter the Great. The top local representative of the bold and brazen IPA, linked to the long-haul sea voyages of colonizing British sailors, was the appropriately named This Must Be the IPA, by Cold Garden. Brewer of the year went to Blind Enthusiasm Brewing, a brewery with a plan to make Edmonton a globally renowned source of traditionally made sour beer, specifically the Belgian lambic. At the time, Blind Enthusiasm was about a year old—an indication of just how quickly the industry was advancing.

Enthusiasm, blind or otherwise, struck me as one factor that brought each of those brewers to the stage. But enthusiasm on its own is like a squad of cheerleaders—they know how to motivate, but not how to win the game. When I met Marc Shields at the awards, however, I began to understand. Under the right conditions, enthusiasm can make a person do crazy things. Like follow your heart. I bumped into Shields near the oyster shuckers, where he was watching the show. He blended in perfectly: black hoodie, dark ball cap, thick beard, friendly and happy to chat. He'd started Siding 14 Brewing, he told us, a brewery in the central Alberta town of Ponoka. At the time, his company was approaching its second birthday.

"This is great for Alberta beer," Shields said of the event. "Great for our beer and for our ingredients."

Those ingredients are particularly important to Shields. A few years ago, he moved with his wife from

Edmonton to Ponoka. He quit a twenty-year job with IKEA, which he still considers to be one of the world's best employers.

Purpose built and opened in 2016 kitty-corner to a tractor sales lot in town, Siding 14 features a bright and tidy taproom with a long bar and a shaded patio near a small crop of hop vines. The taps offer no fewer than half a dozen beers at any given time, each made on-site. Siding 14, like any brewery, was a risk. But the way Shields saw things, its time had come. Or the time for it had come back.

The Alberta Beer Awards were the celebration of an industry, but they were also an acknowledgment of something more communal, a scaling back in a way, at the same time as the businesses were scaling up. Once, Shields explained, every small town had "the four Bs": a baker, barber, butcher, and brewer. Buying local was a necessity. Now, as we saw in the previous chapter, it's a choice he thinks that a growing number of people want to make. Point being, it's not just about consumers, it's also about producers like him. He made his move, back to Ponoka, for the barley. Before he turned pro, Shields was a dedicated homebrewer. Eventually, being close to a brewing supply store just wasn't enough. "I wanted to be near my farmer," he said. "We're coming back to that same concept. Everyone wants to know where their beer comes from." And Shields wanted to know *exactly* where his barley came from.

His move could be seen as a comforting and convenient rationalization from a man who'd bet it all on craft beer, and whose business is built on not just a luxury product, but a luxury product in a niche market. Pushing him forward, however, were the same feelings that other brewers would describe in conversations throughout the awards night, as they explained what brought them to the industry, and why they decided to take a chance.

"I don't want to regret not doing it," said Shields. "Follow your passion—isn't that what everyone wants? I'd rather swing and miss than not swing at all."

When the awards wrapped up later, well before midnight, bartenders swung open fridge doors to disappoint thirsty gala-goers with empty shelves; the Palace had been drunk dry. Nevertheless, the revelry continued. The next day would see many of the attendees back at work to make more of the stuff, but for now they were riding the high of being praised for doing their damnedest to stand out in Canada's toughest liquor retailing environment, one that's flooded with literally thousands of products, most of them backed by the multi-million-dollar marketing budgets. If they were underdogs, this was their night to gather as a pack and howl at the moon.

Earlier, as Marc Shields had watched the last of the awards being handed out, he shrugged and said, without disappointment, "Who knows. Maybe next year, that'll be me."

The future of craft beer in Alberta may be as wildly uncertain as it is wide open, but that night at the Palace suggested that both the traditions and foundations are now settling nicely into place. Whenever Siding 14 may be called up to the stage in the future, Shields would know exactly what to do.

FERMENTING FARTHER AFIELD

As it turned out, Shields did. In 2019, he was back, this time in Edmonton. Whereas he would have faced off against fifty-seven of roughly eighty breweries eligible for the 2018 event, his next go-around pitted him against the seventy-six breweries who entered of the roughly one hundred that were eligible (around four hundred beers were put forward for consideration across nearly thirty categories). When his brewery's name was called to collect a silver medal in the stout category for Coal Pusher, a dark beer noted for its caramel, chocolate, and smoky notes, he and his team went on stage, held the award, hammed it up for pictures, then returned the award, as per instructions. He had indeed been listening.

The award qualified Siding 14 as a standout brewer of the style, but it also made it relatively common in a way that said something important about the nature of the industry's growth. Shields's brewery was part of a healthy contingent of brewers who were succeeding in producing high-quality beer outside of the major centres of Calgary and Edmonton, where the industry is concentrated.

Just as Peter Johnston-Berresford had hoped when he spearheaded the creation of the brewing program at Olds College, and as Shields had predicted a year early with his "four Bs" theory, beer making was taking root outside of the major urban centres as it matured. In fact, at the time of writing, well over a third of Alberta's more than 110 breweries were located outside of its four largest cities, including Red Deer and Lethbridge. Interestingly, they were embracing their status as geographic outliers and finding unique ways to be sustainable, with few taking any cues from their big-city counterparts to do so.

Grande Prairie is not that big, but it's not exactly small, either. It's a city of more than 60,000 people, with a strong sense of economic autonomy drawn from agriculture, forestry, oil, and gas, and even a growing tourism industry thanks to local dinosaur bone beds. It's a hub for a region that includes a population of more than 280,000. With a median age of just under thirty-two, Grande Prairie is one of Canada's youngest cities and also one of its fastest growing. But Grande Prairie is also a long way from almost everyone else in the province, with the nearest major city, Edmonton, some 450 kilometres to the southeast. Set amid the rugged beauty of the boreal forest, it's a kind of urban outpost, a border town along lands unfathomable to the vast majority of Albertans.

"We feel like we're kind of in it alone in Grande Prairie," said Dalen Landis, head brewer and co-owner of Grain Bin Brewing Company. That said, the brewery has made that geographic isolation work to the company's advantage. "In Grande Prairie, everyone either has a small business or they know someone with a small business," said Landis. "So everyone's more than willing to try something new to support a small business."

But Landis and the brewery's five other local co-owners learned early on not to take that for granted. Craft beer production risks becoming a narcissistic exercise should a company of former homebrewers believe that their own tastes should inform the business plan. Once, it was that way for Grain Bin as well.

"When we started, what we wanted to do was make beer that we wanted to drink," Landis said. At the same time, however, "part of us was thinking, 'Maybe we should be a little less selfish with that idea.'" With that, they opened it up to customers with what they called the Democracy series.

Conceived by Landis, the idea was to turn some brewing decisions over to customers. Through polls on the company's website and social media channels, he asked Grande Prairie beer drinkers what they wanted to see in a beer, whether it be style or ingredients. Landis and the team would then apply their own

"Grain Bin flair." They brewed those beers in small batches, the biggest being 750 litres, or about a thousand 650-millilitre bottles and a few kegs for which it would not be "hard to find a home," said Landis.

For the first in the series, launched within the first six months that the company existed, they polled for style. Given the response of "sour," Grain Bin brewed an accessible but tart strawberry-rhubarb fruit beer. Though slowed down by leasing and location issues that required not one but two relocations, the company managed to continue the Democracy series by producing four voter-influenced beers in three years. Each one showed a level of beer awareness and education that might be erroneously considered to be the domain of the big city. A 2018 poll, for example, asked for input on hops varieties. Sorachi Ace, an obscure variety of Japanese derivation, won out. Grain Bin incorporated it into an American wheat ale that drinkers largely applauded.

"With the Democracy series, it's a more sought-after beer just because people are invested in it," said Landis. "[Consumers] want to try it because they're the ones who asked for it."

Though the Democracy series has proven handy as a tool for market research, Landis feels that investment is its most important outcome. "It gives everybody a bit of a sense of ownership because they have a say in something we're manufacturing. I think that has been quite beneficial."

Grain Bin has been able to grow production to roughly twelve thousand litres per month. They have beer on about fifty taps around town, including chains where core beers such as Pipestone Pale Ale and Red Willow Amber Ale offer an alternative (alongside GP Brewing Co., the city's other local craft operation) to lagers mass produced by companies with head offices much farther away than Edmonton. This does not surprise Landis, and not because he has an inflated sense of the quality of Grain Bin beer. Instead, he feels that it's a product of how people are naturally inclined to embrace the places they're from, especially when directly invited to do so. It's yet another variation on the community-building properties of craft beer. Landis hopes to increase the Democracy series run to about three times a year.

"The market is trying to get more local," said Landis. "But I would argue that it's trying to get even more hyper-local. Not only are you going to want to support the brewery in your hometown. You're going to want to support the brewery that's in your subdivision, the one that's on your block."

Demand from outside Grande Prairie will inevitably increase—one of its sours, Grain Bin II, was named best in show at the 2019 Alberta Beer Awards. Anyone outside of the northern city who got their hands on a bottle should feel privileged. Barrel-aged, the beer was one to linger over, replete with the surprising and delightful fruit and funkiness that can come of fermenting with unusual microbial suspects, making it dangerously easy to drink despite coming in at 9 percent alcohol. It's a beer I continued to look for in vain for weeks after finding just one on a forgotten shelf of a grocery chain liquor store. Some bottles do make their way to the cities of the south in more satisfying quantities, but Landis and his colleagues are cautious about undermining the "consumer-driven brewery" initiative that has brought them success at the margins of Alberta's craft beer industry.

"It's hard to send a pallet of beer away when you know that you can sell it in Grande Prairie."

Turner Valley is a lot closer to Calgary than Grande Prairie is to Edmonton, but Jochen Fahr wasn't conscious of any of the potential benefits of operating outside of larger markets when he first saw the southern town. What he *was* conscious of was the feeling that he'd come home. The biochemist-turned-brewer grew up in a village of 300 people in southern Germany, along the foothills of the Alps. An hour southeast of Calgary, Turner Valley is home to fewer than 2,600 people and backs onto Kananaskis Country, that verdant buffer between prairie and mountain.

"It looked exactly like where I grew up," said Fahr, who arrived in Canada in the mid-2000s for school, beginning with a bachelor degree and ending in a doctorate, before eventually ending up in Calgary's biotechnology sector. But the work was not his calling; brewing was, which he didn't see as much different anyway. "It's all biochemical engineering, the whole thing," said Fahr. "People tend to romanticize things. To me, if you take an engineering approach to it you can make a decent beer."

Arriving at craft, however, was a less empirical matter. Beer was in Fahr's blood. Back in Germany, his dad had worked for a couple of brewers and sometimes young Jochen would tag along. Eventually, he took to making his own. In Canada, he won awards for his traditional German beers in competitions in Calgary, Edmonton, Halifax, and more. One year in a Vancouver competition he won best in show for his hefeweizen—"not an IPA, not a porter," Fahr emphasized to me. He used the recipe to make the beer for his own wedding.

When he'd had enough of the biotech sector, beer was there for him. In January 2015 he packed up his office and in February incorporated Brauerei Fahr. While he made his first beers at other breweries in Calgary and Edmonton, he wasn't sure either city was the right place for him to set up shop. He wondered if there might be another option where he could be slightly removed and let the rapidly accumulating breweries in Alberta's bigger centres "step on each other's toes," he said. That's when he remembered Turner Valley, a place he'd once passed through and that was only too happy to have him, going so far as to source a 5,400-square-foot facility for him, complete with a two-inch water line, "which is important for a brewery." There, he could get on with his plan: Brew in a way that Germans had brewed for centuries, which was how he felt very few Alberta brewers were brewing.

Fahr is a traditionalist. His core beers, which he produces in a large, thirty-barrel brewhouse, include his prize-winning hefeweizen, a crisp pilsner, and a dark wheat beer. He diverges to make a few seasonal brews but endeavours to keep things simple, adhering as closely as possible to the Reinheitsgebot, or German purity law. Decreed in 1516, it prohibited any beer from containing any ingredients other than barley, hops, and water, with yeast excluded until its discovery a few centuries later. (He acknowledged that hefeweizen, made with wheat, wouldn't make the cut but, even for a pragmatist like Fahr, some rules were made to be broken.)

Even his packaging speaks to a minimalist tendency, deliberately avoiding anything overtly artistic that might contribute to what he feels is the Jackson Pollock–like chaos of the average retail beer shelf. In contrast, his labels feature nothing more than a plain background with the company logo and as little text as possible. What's more, they're plastered onto glass bottles rather than the aluminum cans that contain the vast majority of Alberta craft beer. Fahr insists that conventional brown, capped bottles do a better job of excluding the oxygen and UV light that can lead to spoilage.

He sees his contrarian approach as an advantage. "I have my own niche," said Fahr, who was not yet forty when I interviewed him. "If you make traditional beers in good quality and quantity, that's a big part of innovation."

Jason van Rassel, Edmonton-based beer columnist and blogger at originallevity.ca, sums Fahr's work up this way: "In an age when people are doing milkshake IPAs flavoured with breakfast cereal, he's decided to brew German-style beers. This is just one geek's opinion, but he's one of my favourite brewers because he does those styles and he does them really well."

That breakfast-cereal IPA (no joke, they're out there) underlies Fahr's criticism of what he saw as the direction of Alberta craft beer at the time. The industry's prevailing impulse, he suggested, is to complicate styles and follow trends that ebb and flow.

"It's great that people are playing around with recipes," he said, slightly worried about coming across as curmudgeonly. "I feel like I'm beginning to be a balance point to that." He'd rather be an alternative to the alternative that is modern craft beer, and offer easy-drinking, high-quality lagers and ales—entry points for those who may not be ready to commit to the extremes of craft.

That said, Fahr sees plenty of opportunities to occasionally challenge tastes with tradition. Among his core beers is a pilsner made in a style that dates back centuries. In the region of Germany that he once called home, it's made through decoction, or boiling a portion of the mash, the step during which starches in barley are naturally converted into sugar for fermenting, before reintroducing it to the process.

"You build more body," said Fahr. "It's very traditional. If I wanted to be ultra-traditional, I would do that three times."

Practical and mindful of his market, and perhaps of the role he feels he should play in it, he holds back. One decoction will do. Any more than that, and "I feel it's too complex for the market."

The small-town trend with craft beer continues in Banff, the emphasis on small. Among the visitors who venture behind the scenes at the Banff Ave Brewing Co., the most common reaction, said Miranda Batterink, is shock.

"Shock at the size of it—the *small* size of it," clarified the pub's head brewer.

Opened in 2010 as part of a family of breweries that includes outposts in Jasper, Calgary, and Edmonton, Banff Ave is the only brewery in Banff. For a town of its size, with just under eight thousand residents, that would be more than adequate. But Banff is not your average small town. Nestled in the Bow Valley of the Rocky Mountains, it saw more than four million visitors over the 2017–18 season. Not every one of those visitors arrives demanding a local craft beer, but enough of them do that Batterink is still amazed every now and then at who's drinking her beer.

"Sometimes I'll be in the pub on a busy day and look around and be like, 'There are probably thirty-five different countries here right now,'" she told me one day between batches. And then, of course, there are the locals, too. Being amazed may be a good way to keep from being overwhelmed. In the summer, Banff is probably Alberta's most densely populated place, and Batterink has to work hard to meet demand on a tiny, fifteen-barrel brewhouse. Every day is an exercise in resourcefulness, "juggling tanks" to move through as much beer as possible.

"You don't really have a day to screw up," she said. Planning is the key. "You can't speed it up. You don't get to decide when your beer finishes. Your beer will tell you that. The trick... for us is predicting what are we going to run out of two or three weeks down the road."

The other catch is that she has to do it in an ecologically friendly way. The Town of Banff draws its water from underground wells and takes conservation seriously, with bylaws that guard against excess use for commercial purposes. This doesn't quite wash with common brewing practices. In the US, breweries use an average of seven gallons of fresh water for every gallon of beer produced (roughly twenty-six litres for every four litres of beer). This is more water than required to produce oil from Alberta's oil sands, where the ratio is about three to one. To lessen the strain on the local supply, Batterink's juggling borders on acrobatic.

After the wort, beer's precursor, is boiled in a tank called a kettle, it must be rapidly cooled. If not, high temperatures kill the yeast, a tragedy in that the wort does not become beer. A heat exchanger does the trick, using water as coolant before the microbes are added. At Banff Ave, the size of the brewery challenges this process. Batterink can't just run the cooling water down the drain without raising the ire of town officials, but she doesn't have a spare tank to put it in, either. The result is a lot of extra work that city brewers don't have to do. Batterink drains that cooling water into another tank called the mash tun. Before that happens, the vessel has to be emptied and cleaned of the grain used to make the wort, the stuff that had gone on to the kettle. Sufficiently cooled, the wort can then go from kettle to fermenter, where the magic (or, if you're Jochen Fahr, biochemistry) happens. The kettle then takes on a new job: heating up that cooling water so that it can be used to make another batch of beer. For that, another transfer ensues, this time from mash tun to kettle, where that cooling water can be warmed so it can be transferred *back* to the mash tun the next day and ultimately become a new batch of beer, and the carefully choreographed process continues.

It's time-consuming but necessary if Banff Ave is to be a decent corporate citizen dedicated to preserving one of the world's most beautiful places, as well as supporting the community and those who live there year-round. The brewery contributes to the town and surrounding Bow Valley area with quarterly brews that share proceeds with local charities and non-profit organizations. That's all for nothing if companies, especially those as potentially resource intensive as breweries, play fast and loose with the necessities of life.

For Batterink, who was drawn to the area from her home in Ontario for the hiking, climbing, and biking opportunities the Rockies offer, conservation efforts align with the ethic of craft beer itself. Local brewing is not just about making sure you can meet demand; it's about how you meet that demand. Naturally, Batterink wants to see her beer sell. But she does not want to contribute to the possibility that patrons might drink the place dry. Of water, that is. "Craft beer by its nature is very locally driven," she said. "You're more accountable to customers."

NICHES WITHIN A NICHE

If there is an admirable small-town element to Alberta's craft beer industry, there is also an equally admirable small-business component. Alberta craft beer is far from being a mature industry (more on that later). But it's already begun to show signs of being able to support ancillary businesses that would not have otherwise existed in the province. These go beyond the usual suspects of craft brewery tour companies or beer subscription clubs. Instead, they're being led by entrepreneurs who've recognized a need in the industry that might make a true contribution to its longevity, and who have brought unique skills to the job of meeting it.

Brian Smith had a sense that the marketing people might not be very happy if they heard the way he was talking about how Wild Rose Brewery had been putting such skills to work in brewing its IPA as of late. Instead of adding the industry-standard pelletized form of hops, in this case Columbus, an intensely aromatic but mildly citrusy variety, they'd done away with the plant material itself. All they wanted was its essence: those oils that give beer its pleasant bitterness, as well as the bulk of its flavour and aroma. That essence, though, isn't exactly pretty.

"It's almost like a paste," said Smith, director of brewery operations at Wild Rose, a veteran of Alberta craft beer (as well as the province's first to be bought by a multinational; a few weeks after my conversation with Smith, Sleeman Breweries, owned by Sapporo, purchased the Calgary brewery).

He was talking about the hops extract Wild Rose had begun using in its six-thousand-litre batches of a beer that they'd been making pretty much since they started up in 1996 in a Quonset hut at the old Currie Barracks in southwest Calgary.

"It's a fairly thick, viscous substance that we have to heat up a little bit to get into the brew kettle," said

Smith. "It's a dark green—[it] almost looks like motor oil." But, he added, "It smells delicious."

The economic viability of that IPA was once questionable. In its early days, the beer sold so poorly in winter that it was put on hiatus until warmer months. Now, however, it's a mainstay for Wild Rose, and may even rank as one of the province's most efficiently made brews. That dark green "paste" is to thank for it—along with Diana Powers.

With her brother, Powers owns and operates Aratinga Extracts, a commercial lab in northeast Calgary. She specializes in preparing botanical extracts, a field she began exploring as an undergraduate in her native country of Colombia before earning a scholarship at the University of Calgary, where she took a masters and a doctorate in chemical engineering. For years after, she was a researcher at a pilot upgrading plant for a major oil company.

"Something in my heart was telling me there was something else," Powers said. Eventually, she followed it. With her family's support, it led her to explore entrepreneurship in the subject that attracted her to chemistry in the first place. The timing was right, since, all around the city, a new industry was emerging and filled with business owners who might be looking for any advantage they could get.

"With the boom of the craft brewing industry, we realized we could do hop extracts," said Powers. She began cold-calling breweries, asking for meetings.

Particularly in the United States, a growing number of craft breweries are turning to extracts to improve processes, yields, and bottom lines, without compromising flavour. The practice is gaining ground in Canada as well, at least as a concept. Peter Johnston-Berresford has even seen interest in it among students at Olds College who typically cherish the more hands-on, artisanal aspects of craft brewing.

"I credit them for being so open-minded," he said. Once, he might have expected outright resistance, something along the lines of, "'You're an embarrassment to the trade,' or 'That's not craft.' Now, they say, 'That's a good idea, because it's much more efficient, much more fiscally responsible, and it creates batch consistency.'"

The other reason for the growing receptiveness to the use of hops extracts may owe to the extraction process itself. Powers stressed that deriving hops oils is a physical process, not a chemical one. She places pelletized hops (dried and compressed plant material) in a closed vessel, brings the temperature to about 40°C, and pumps in carbon dioxide to raise the pressure to four thousand pounds per square inch. "At those conditions, the CO_2 becomes a very good solvent," said Powers. The result is that "paste" whose bittering qualities can be quantified and verified before it's sent back to the brewer.

Smith noted that the extract leads to a cleaner brew, free of a large portion of "green matter" that can come of using hops pellets. But the benefits may go further. A pellet acts as a sponge. As a batch of soon-to-be beer boils and flavouring oils and acids are released from the hops, those pellets absorb liquid. Brewers don't get that back. Instead, it gets chucked out with a mushy mass of pulp that makes Smith's description of the alternative sound pretty appealing. Ultimately, that means less beer and fewer profits.

Early numbers showed improved yields at Wild Rose, said Smith, suggesting that the investment in the extract could lead to better returns.

"As things get more competitive, you're definitely looking for things that can help you both from a quality and a bottom-line point of view," said Smith. He knows that, from the perspective of consumers, innovation in the industry is often seen as "just a million different styles of beer." But what's the point if a producer breaks the bank making them? "At the end of the day, everybody's got to do what they can to make a living. So, anything that we can do to improve our product, improve our process, that's something that we're open to."

Across the Elbow River from Wild Rose, in a light industrial park in the southeast, Prairie Dog Brewing is open to that as well. Like Wild Rose, they were using Powers's extract enthusiastically but judiciously, crunching numbers and checking flavour profiles.

"We're in the best position to take advantage of this," said Gerad Coles, one of a group of founders who'd packed in tech jobs in Silicon Valley to come home and start the brewpub. Their idea was to bring southern-style barbecue, which was big in the Valley, to Calgary and pair it with craft beer made on-site in a ten-barrel brewhouse, 1,000 to 1,200 litres at a time. They opened their doors in June 2018 and sell every drop exclusively on location at retail rather than wholesale prices. That means more profit per pint.

Coles felt they could improve on that profit. At the end of a brew, he'd be left with a couple of five-gallon pails of spent, spongy hops. "That's a lot of hop material that's sucking up liquid. If that wasn't there, we could get higher efficiency for the whole brewing system. Ten gallons of beer is worth a lot."

He'd heard about extraction being used in the cannabis industry, where it is applied to purifying cannabidiol, the compound responsible for many of the drug's therapeutic properties. But, despite legalization in October 2018, he hadn't heard of any local companies doing the work, let alone applying it to hops. Then Powers contacted him.

"It's not something I would have imagined coming to Canada at all, and especially not to Calgary, for a long time," said Coles. "I was really stoked to see that Diana was taking a plunge into this market."

When I spoke with him, he was still in the testing phase, with plans to switch out the Lemon Drop hops in Prairie Dog's Tail-Twitcher IPA. He was taking a gradual approach, using the extract along with pellets, so as not to introduce too many changes at once. But he felt confident that it would produce the results he was after, which involved more than just making good beer.

"If you have more product coming out of the bright tank, that's either a reduction of labour because you have to brew less often or it's more product that you can put into the market and earn revenue off of. Both of those things are great for business."

By spring 2019, Powers had been seeing more of that attitude in the brewers she contacted than a year earlier, when responses were more hesitant. Though she hopes one day to expand her business to prepare extracts from other botanicals, including cannabis, she would like to reach capacity, and the point where she can scale up, with brewers.

As a small-business owner, and a recent startup as well, that prospect of Aratinga's success also appealed to Coles. He liked the idea of being a catalyst for economic diversification beyond beer. "The growth of the craft beer industry here is hopefully going to lead to dozens or even hundreds of businesses that are supporting it in all these other ways," he said. "I feel like there's a positive effect of spending money on these other local businesses and supporting them in their growth. So even in the short term if [using extract] doesn't result in higher profits, I'd still rather spend that money on a local company and help them build up their skills so they can work on their own efficiency improvements and maybe come out with another product based on their learnings with us that adds even more value. We can strengthen our economy through beer and all the businesses that it spawns."

Like many new brewers, Powers felt she had much to learn about building a business. When we talked, she'd been a full-time entrepreneur for roughly two months. She believed she'd need about four consistent clients with needs along the lines of Wild Rose to max out her machine. Though some were showing interest, they'd yet to commit. But that didn't seem to worry her. In a way, she'd already accomplished what she set out to do. Powers liked her old job, working as a chemist for an oil company, but she could no longer deny "the vision of doing my thing to help other people," she said. "It was painful for me."

Craft beer, or at least the hops, took the pain away. "Now that I'm doing this," she said, "I feel pretty happy."

INSTEAD OF FINDING a niche no one had yet thought to fill, as Powers did, the Hamill brothers decided to find a space inside a near monopoly in Alberta. There are two companies in the province that provide local malt: Rahr Malting in Alix and Canada Malting in Calgary. Or, rather, there *were* only two companies. Make that four. In Strathmore, fifty kilometres east of Calgary, a family farm provides the source barley for Origin Malting & Brewing. At Red Shed Malting, however, the Hamills have taken that concept a step further. And they felt it was worth a trip to Turkey to make that happen.

Because despite the hundreds of thousands of tonnes of malt produced by Rahr and Canada Malting every year—malt that's prized the world over—there was still a gap. On the family farm twenty minutes south of Red Deer and just off Highway 2, Joe and Matt Hamill were the first entrants into Alberta's craft malting business and the first to be able to directly link a brewer to a particular farmer.

Like craft brewing, the operation involves small batches fretted over by few people, but with even clearer origins of ingredients than the brewers themselves can account for. At Rahr, bins of finished malt have been sorted by type and quality, not by farm. They know which dozen or so farms each batch came from but preserving the identity of single kernels would not be practical, let alone cost effective.

At Red Shed, barley can be harvested from the Hamill family's two thousand acres, or any other farm, and augered directly into an immersion tank, that first step in malting. There, Joe, a prodigiously bearded, sinewy, early-thirtysomething, starts on the process using an amount of grain that would barely make a splash in the sea of malt running through Rahr each day. But producing base malt was never the main objective, said Joe. Rahr and Canada Malting do that just fine. When the brothers first became interested in finding a way into the craft beer boom, they realized that, while there was plenty of base malt being produced, nobody local was making specialty malts. If base malt is a kind of skeleton, specialty malts flesh that skeleton out, and are used to make a beer not just in a certain style (be it anything from an IPA to a darker-than-midnight stout) but unique within that style. They're what make it possible for a brewer to personalize her or his product and contribute to endless permutations of recipes. Those specialty malts are created with a malt roaster. The Hamills were the first in Canada to have one.

At first the brothers wrestled with what seemed like two paths into craft beer that were competing for their attention: brewing or malting. Joe was a homebrewer who wanted to use malt made from the family's own grain in his beer. (Truth be told, Joe still brews, and in the corner of the malt house is a tiny, two-hectolitre brewing system. Occasionally, batches are sold under the banner of Hamill Brothers Brewing.) His brother, Matt, older by about four years, also felt that connection to the land as he considered the options. "Malting is more true to our background," Matt said. "Dad has the farm. Malting was the next extension."

It's a family affair. Matt and Joe's mother, Susie, does the accounting. Like the brothers, she's part owner, as are Joe's wife, Daelyn, and Matt and Joe's father, John. John was the original owner of the eponymous red shed, which was nothing more than tin walls and a gravel floor before the boys took over half of it and made it habitable. To keep the dream moving toward reality, the brothers took two weeks of vacation—Matt from his job in banking, Joe from one in drafting—and headed to the Canadian Malting Barley Technical Centre in Winnipeg, Manitoba, to take a two-week course at the Malt Academy. "Yes, that's what it's called," Matt said when I raised an eyebrow to suggest I thought he was joking.

After that, they began to source equipment. Joe and his dad went to China to look at the immersion tank and two malting boxes (each the size of the business end of a decent-sized dump truck) that they ultimately bought. But the real score was in Izmir, a city on Turkey's Aegean coast. It's also home to Toper, a company that makes coffee roasting equipment and was willing to do a malt roaster on spec. Clearly, however, it was not Toper's core business. The brothers recalled seeing how neatly and carefully the new coffee roasters were crated for shipping from Turkey. That wasn't the case for the malt roaster.

When it arrived at the farm, "We open up the sea can and it's just saran wrapped and bolted down," Matt recalled. They drove in a forklift and gingerly began to try to move it.

"[Toper] said it weighed so much," said Joe. "It probably weighed twice as much as they said it did."

They weighed down the back of the forklift as a counterbalance, but the load still teetered precariously as they transferred it across the yard to its new home in the malt house.

"Our forklift was too small," said Matt. "If that would have tipped over, and it was damn close, that would have been the end of things before they started."

Officially, that start was October 2015, when Red Shed produced its first test batches of roasted malt, a process that takes anywhere from forty-five minutes to three hours (after malting) depending on the desired toastiness. These days, the company has a roster of around twenty clients and Joe is malting and roasting full time. He also moved back to the farm, something that always interested him but that he once thought impossible. The farm, as it was, was simply not a big enough business to support multiple families. Now it is, and Joe's commute is a few minutes' walk from a new home he built and lives in with his wife and a toddler, with another one expected soon after our interview.

Matt, too, hoped to get back to the farm and was making plans to move home. His plan was to take up residence in the "homestead"—the original house built by his great-grandfather, who immigrated to Canada in 1929 from Ireland. After crunching the numbers, he believed that the company was on track to profitability before 2020 (though he suspected that having to start paying himself might mess with the math). Growth, however, was measurable in increases of several hundred percent per year, though Matt admitted that was also a factor of recently starting from zero. Still, the Hamills were quickly making themselves essential to the brewing community. Matt, for example, was known for helping brewers not miss critical brew days with "ridiculous timelines" by personally delivering specialty malts as needed. But Red Shed was also playing a role in connecting brewers directly to the farms they trusted and wanted to support.

Interest in that connection is on the rise. Barley from the farm that supplies Siding 14, for instance, comes to Red Shed for malting. A beer called Bock Chain, produced by Calgary brewpub Last Best, was the culmination of tracking barley from a specific section of land, then to Canada Malting, then to Red Shed for roasting, then back to the city for brewing. But there is probably no better, nor extreme, example of what the craft maltster makes possible than Blindman Brewing's 24-2 Stock Ale.

Blindman Brewing co-owner and marketing head Kirk Zembal pointed out to me that Red Shed is in almost every beer the brewery makes. It's an accessible, local alternative to product that would often otherwise have to be sourced from out of the country. But the ale, an old English style, was unique. The brewery partnered with Lacombe-based 24-2 Draft Horses to plant and later harvest barley using old ways and equipment. The idea was to explore the often-overlooked link between brewers and the land, but also to dig, if you will, a little deeper. Zembal told the story in several small-print paragraphs on the side of each can of beer, shining a light on the relationships. The "12 powerful Percherons" got their due, but so did Red Shed, without whom the dark, malty beer could not have happened. "It's fantastic to have such a great maltster in the neighbourhood to work on custom malts and share their craft with us," Zembal wrote.

The key to that ale is a traditional brown malt. Rather than remove the majority of the moisture from it through kilning, it was transferred wet to the roaster, which almost caramelized it, said Joe. It was a learning experience, but not always a pleasant one. The wet malt clogged hoppers, making for slow going at several stages in the process.

"We wouldn't be able to do a bunch of really cool stuff without them," said Zembal. That "cool stuff" goes beyond the stock ale, which Blindman hopes to do every year with Red Shed's help. The brewery has continued to ask the Hamills to push the limits of its roaster, and possibly put it to work processing ingredients such as malted oats and triticale, a wheat-rye hybrid that grows well in Alberta and that Blindman has begun using.

"Red Shed really allows us to fully innovate in all aspects of brewing—and not just in putting together a bunch of commercial malts to create a recipe. It allows us to craft the malts to our desire, to more fully realize an idea." What's more, Zembal added, thanks to the nearness of roaster and brewhouse, every batch is fresh. "Freshness in malt matters a lot in flavour development."

That willingness to tackle the learning curve, to take a chance on malting the way that a brewer might be willing to take a chance on a beer, may be one of the factors that could make Red Shed the success the brothers hope it to be.

"I'd like for it to eventually flip around so it's the malt company helping out the farm, rather than the farm helping out the malt company," says Matt.

Until then, the roaster is much less precariously positioned than it was when it arrived, untested and unwieldy. While it's not likely to ever eat into the market share enjoyed by Rahr and Canada Malting, it may even prove to have an advantage nationally. "If you want to do a dark beer and you want to use only Canadian ingredients, you've got to use our stuff," said Matt.

Which is a pretty good niche to occupy.

CO-OPETITION

Almost from the start, and certainly following deregulation, Alberta craft beer has been an industry in which its members knew that they needed each other. Beer making in the province was the business of multinational corporations who not only dominated the market, but had begun to compete with craft by, in a sense, masquerading as craft. There is no shortage today of veteran and highly respected small-batch breweries that have been scooped up by Big Beer. Lagunitas Brewing Company, for example, is owned by Heineken; Molson Coors owns the likes of Granville Island and Creemore Springs breweries; and, of course, Wild Rose is now under the wing of Sleeman, in turn owned by Sapporo.

The reason for those acquisitions is another complicating factor. The market for beer, in general, isn't currently growing. The only way for new brewers to succeed was to try to steal market share. And the best way to do that was to present a united front and leverage the strength of even relatively small numbers. It is something of an enlightened state: an immature industry that's able to take a mature approach to working together to achieve a common goal. Or maybe it's a bunch of underdogs who recognize that they're going to have to scavenge for scraps unless they team up to take down big game. Either way, the results are the same.

There are few people in the Alberta craft beer industry who promote and defend the persisting ideals of collaboration and community among members as staunchly as Shane Groendahl. At times, that has involved getting tough. At the 2019 Alberta Craft Brewing Convention in early March, Groendahl led a session to organize the next iteration of what's known as the Unity Brew. Each year since October 2009, when the first event was held at Canmore's Grizzly Paw Brewing Company, the Unity Brew has brought industry members together to brew.

"Unity Brew is one of the most anticipated brewer-focused events of the year," Alberta Small Brewers Association marketing and events manager Lauren Reid told me. "[It] offers a unique opportunity to get everyone together and facilitate the creative and collaborative side of this industry."

But the Unity Brew has become much more important to the local craft beer community than simply providing the chance for a recharge. That's why, that day at the convention, Groendahl took his job so seriously, and handled it with a sternness that's not often seen in an industry that prides itself on having struck a balance between good times and hard work.

Maybe it was the samples. To help stimulate discussion, Groendahl had brought a flat of the last year's Unity batch, a red ale that put Alberta malts, and therefore maltsters, front and centre. Set at the back of a small room at the Edmonton Convention Centre, it was nearly empty before Groendahl had even got to the podium to say it was there. The problem was, not everyone who'd helped themselves returned to their seats to sip and listen. Some gathered near the door of the room to chat noisily. It competed with Groendahl as he began the session. He paused.

Groendahl, it should be noted, is not an imposing young man. He is of average height and considerably slimmer than almost any brewer you will ever meet (and he's invariably clean-shaven). He is polite, with a businesslike manner, suffers no fools, but does not take himself seriously in any off-putting way. That said, he is often serious. It's easy to take him seriously because of that.

Standing at the podium, Groendahl pointed to the group near the door. The room went silent—except for the gang standing around the door.

"If you're conversing," he said firmly but not angrily into the mic, "get out." They did. Groendahl moved on without comment, like a judge who'd just lowered the gavel. The business at hand was too important, there was too much to be done, and people were depending on him.

The Unity Brew began as a community-building event, and it still is. Now, however, the impact is measurable, because it has become a key fundraising tool.

In 2018, it generated $40,000 for the Brewers Association, which represents more than 80 of the province's 110-plus small breweries as a single voice to work with government on policy, the AGLC on operations, and whoever else might play a role in advancing the state of the industry. In essence, it's a formalized extension of the co-operation that goes on informally among breweries in the province, whether they're association members or not.

"We like to use the term 'co-opetition' in the industry," said Barb Feit. The former Big Rock CFO helped found the association in 2013. "Everyone is technically a competitor but the collaboration in how the industry comes together, not just in making beers, is refreshing."

Indeed, collaborative beers are common among Alberta brewers. To Feit's point, however, the cooperation extends beyond shared products. Tool Shed Brewing co-founder Graham Sherman once told me that staff at the Dandy Brewing Company would come to use his facilities to wash kegs. Adam Corsaut, an owner at Edmonton's Analog Brewing, mentioned buying equipment from Alley Kat Brewing, just a few blocks away, at prices one might think were reserved for friends and family. And then there's Groendahl's side project: Edmonton Beer Geeks Anonymous.

Started in 2011 as a series of educational sessions and small events like beer-and-movie nights, EBGA has evolved into a showcase for Alberta craft beer. Each winter it stages Freeze Your Cask Off!, featuring more than a dozen "pins," or twenty-litre metal vessels, from Alberta brewers for about 150 patrons. In the summer, Groendahl and the rest of the Geeks (he insists he is only the group's de facto spokesperson, not its leader) hold the Real Ale Festival. For the 2019 event, he hoped for sixty casks—or, more specifically, forty-litre "firkins"—and about 550 attendees. The events always sell out and are designed to bring together a community of craft beer lovers (the Geeks' slogan is "The search for good beer need not be lonely"). They're also meant to give rookies exposure. "We try to feature a new brewery whenever we can and give them a leg up," said Groendahl. "We sought to give them a soapbox in front of a crowd that was enthusiastic and wanted . . . to see what new breweries could do."

Groendahl does this work for other breweries while running one of his own, as part owner of Blindman Brewing. He sees no issue with providing, in essence, free marketing for his competitors. In any other industry, this would likely be considered a bad business practice.

But there's an anthropological precedent to suggest that it's not, even that it's natural. The fittest survive if others do, too, and the human species is among the best examples of this. Research has shown that humans have been interdependent pretty much since we learned to walk upright. Early humans contributed what they could to group success and collectively persisted in the face of terrible weather, terrifying predators, and lack of craft beer. Later, they'd band together in much the same way to face Big Beer. A common enemy made everyone stronger if they worked together.

That seems to come naturally to Groendahl. He may take things seriously but he doesn't tend to see all those things as work, whether it's helping to organize a cask festival or rounding up donations of malt and canning materials to keep costs down on the Unity

Brew. He'll admit that he's got a lot going on, but he won't say that he's got too much to do.

"Beer is 85 to 90 percent of what I put my energy towards," he said. "It's so much fun. I do what I do because I like doing it."

Back at the 2019 session, he got down to the agenda. Over the coming weeks, they'd propose styles for the Unity Brew then whittle those down to a short-list for voting. They'd need to pick a date and stick to it. They'd need a committee of about seven people. And a venue. Blindman hosted last year, said Groendahl. Who'd volunteer their brewery this year?

Two or three hands went up.

He warned them that it would take some effort, and that no "take-backsies" would be tolerated. He also reminded them that there was a point to it all. The Unity Brew is "not just a piss-up. It's not just a means for brewers to get drunk. Although that has happened before."

One of the reasons for getting everyone together was, just as it always had been, to have some fun, but Groendahl wasn't about to take that lightly. The group depended on it.

A GOOD HARD LOOK AT ITSELF

Maturity and objectivity go hand in hand. Nobody grows unless you can take that hard look at yourself, see the flaws for what they are, and then be willing to do the hard work required to make a change. But it takes some growing up to get to that point, where a balance has been struck between confidence and humility. If Alberta craft beer is any indication, industries are like that as well. There is good beer here; the awards hanging on brewery walls across the province prove it. But there is room for improvement. The good thing is that we're willing to say as much. In fact, right in the middle of that exponential curve in the growing number of craft breweries, one critic did just that.

In 2016, beer columnist and blogger Jason van Rassel wrote a post for a Calgary-based blog called dailybeer.ca. He intended it as "a shot across the bow" for the province's nascent craft beer industry. Accolades should not include badges for participation, suggested van Rassel; breweries, new or established, should know what they were doing.

"What I was seeing was a lot of breweries [that were] sorting out basic competencies after they were opened and expecting the public to be guinea pigs," he said. He recalled a local wit beer, a kind of wheat ale, that tasted distinctly but inexplicably of cigarette butts. Did the brewer taste it before packaging and decide to try his luck anyway? van Rassel wondered. Was the consumer not being given the credit it deserved? Was excitement over the new industry compromising objectivity? Whatever the case, van Rassel believed it was time to stop applauding breweries just for hanging out their shingle, and he said as much.

Owen Kirkaldy, one of Canada's few beer judges certified at the master level and co-founder of the Alberta Beer Awards, was even more frank in his assessment. "There are some shit breweries in this province," he said during a conversation we had about the state of the industry and the genesis of the awards. He did not say which.

These, he was careful to point out, represented a minority. Once an award-winning homebrewer, Kirkaldy has hung up his mash paddle. "I stopped brewing three or four years ago because all of the reasons I started brewing have gone away," he said, confident that any style he's after is being made, and made well, by Alberta brewers. In that sense, the spectrum of quality available in the province is similar to that of the world's most revered beer jurisdictions. The reputation of California craft beer in Alberta, for example, is based on what an importer brings into the province, which naturally will be the best beers of the bunch.

"You're not getting Eddie's micro-brewery on the corner where they don't wash their hands," Kirkaldy said. "Not every beer in Alberta is a superior beer, but the best beers in Alberta are right up there with the best beers anywhere."

Also important to note, said van Rassel, is "the speed with which we're catching up to more mature craft beer markets." In part, he attributed this to the growth in brewing talent that comes with each graduating class of the brewing program at Olds College. Good breweries have been started by home-brewers who have successfully scaled up, he said, but at the same time "that program is sending out all kinds of qualified brewers who are doing a really good job of filling those spaces in up-and-coming breweries in Alberta."

Another factor is that brewers themselves seem to have a genuine desire to make a better product. Van Rassel pointed out that he had noticed an encouraging trend in breweries actively involving drinkers in their processes.

When I first encountered this, I thought it strange. Upon release of its Power Up Porter, Edmonton's Analog Brewing labelled the can a "beta" version. It fit with the brewery's video-game theme, but it was also sincere. They wanted drinkers' opinions, and they made it clear on the can that the nature of its contents was up for discussion.

Van Rassel did not think this strange at all. "We have to recognize that beer is a product that lends itself to tweaking." Brewers, he continued, are creative people who, like an artist unwilling to consider a painting complete and who keeps dipping the brush in the palette, will keep refining. What made Analog's request all the more interesting is that the porter, regardless of being of "beta"-version quality, earned a gold medal at the 2019 Alberta Beer Awards.

Those awards are possibly the best example of the

industry having arrived at the ability, and willingness, to take a good hard look at itself, even if it's masked by being one of the best parties of the year.

The Alberta Beer Awards are the brainchild of Kirkaldy and fellow judge Jason Foster, creator of onbeer.org. While judging a competition elsewhere in 2016, they got to chatting about Alberta beer and came up with a plan for an exclusive competition, which they then took to the Alberta Small Brewers Association. The association loved the idea. But what they loved most about it was the opportunity they saw for brewers. The association wanted to use the event to educate; as much as they wanted to celebrate the talent in the industry, they wanted to develop it. If a brewer didn't come away with a medal, they should leave with knowledge. If Kirkaldy and Foster could make sure that every brewer received feedback on beer they submitted, the association would give their blessing. And they would run the thing.

Foster and Kirkaldy were happy to oblige if it meant helping to grow an industry that was becoming highly capable of keeping them supplied with exceptional local beer. It all came to pass as planned. Every submission now receives detailed feedback, win or lose. In 2019, about four hundred of those documents were generated. Be it praise, criticism, or both, the pair endeavour to make it constructive.

"These awards are not about pointing and casting

shame and saying, 'You're garbage,'" Kirkaldy said. "That doesn't help anybody as far as I'm concerned."

As Foster points out, there would not be much opportunity for that kind of approach anyway. The 2019 awards reassured him about the future of Alberta craft beer.

"That there's such a wide range of beer that won [awards] has me quite heartened about the state of the [Alberta] beer industry," said Foster. "We may have more than a hundred breweries, but if forty of those can win a medal, that means that a lot of men and women make pretty good beer out there. That really speaks volumes that we're not just growing in quantity. We're also growing in quality."

The diversity of the entries, added Kirkaldy, was now forcing the event itself to play catch-up in terms of its ability to recognize the quality, adventurousness, and creativity of beer in Alberta. Kirkaldy said to watch for the "barrel-aged beer" category to be split into three streams in 2020: barrel-aged sours, barrel-aged with fruit, and conventional barrel-aged brews. One of the biggest reasons for this was that obscure, high-test beer made by Grain Bin Brewing, the deliciously tart Grain Bin 11.

"Who was making sour barrel-aged beers four years ago in Alberta?" asked Kirkaldy. "Nobody."

5

HERE TO STAY

This boom goes on—and on?

DEBUNKING THE BUBBLE

Of everyone I interviewed, Don Tse was by far the most optimistic about the Alberta craft beer industry. It wasn't just because a rising local supply would help him increase his personal tasting tally. When I spoke to him in the spring of 2019, the Calgary-based consultant was at more than twenty-one thousand distinct beers. "Most of those would be samples," he assured me. "I live a fairly disciplined life in terms of diet and exercise."

Instead, Tse based his positive prognostication on twenty years of watching the Alberta beer scene evolve (he's earned the nickname the Don of Beer). In addition to consulting, he writes, imports and exports, and even co-hosts events like the Alberta Beer Awards, giving him plenty of insider experience and insight. But if he's right about his industry projections, maintaining that lifestyle discipline could become more of a challenge than ever. Given the rapidity of the growth in Alberta's craft beer since 2013, with the number of local breweries doubling between 2016 and 2018, and increasing five-fold since deregulation, I wondered if the province was experiencing a beer bubble rather than a boom.

In bubbles throughout the ages, dating from the Dutch "tulip mania" that inflated bulb prices in Holland in the 1630s, to the US housing crash of 2008, the trouble has always been that investor speculation drives up prices. That then spins itself into a positive feedback loop divorced from reality. Speculation drives speculation, and the value of an asset artificially skyrockets. When you look at the stages of a bubble, as described by economist Hyman P. Minsky, there are craft beer parallels. Stage one—displacement—sees an event (such as the removal of the minimum production requirement in late 2013) capture the interest of investors. Stage two—the boom—sees an increasing number of participants enter the market due to media coverage, speculation, and fear of missing out. In stage three—euphoria—caution is thrown to the wind as the players get caught up in an excitement that creates a disparity between what the market will tolerate and prices (or the number of breweries, in our example). Stage four—profit-taking—comes when the smart money exits early. Stage five—panic—results when the remaining players fail to liquidate (ha ha) in time. The reverberations are the popping of the bubble when it's over and beer's only value is as a receptacle for tears.

That's the worst-case scenario, and Tse does not think it will play out that way. For him, the only bubbles in the industry are the ones rising rapidly and joyously to the surface of every freshly poured pint of lovingly made, small-batch Alberta beer. The reason for his perspective is, in many ways, Portland. Since the first brewpub opened in Oregon's largest city (population, roughly 650,000) in 1985, Portland has become a mecca of North American, if not global, craft beer. It's not the American city with the most craft breweries today, a title that now belongs to Chicago. But with just over 110 local beer makers in the metro area, it has as many breweries as Alberta.

"If we had the same number of breweries per capita as Portland," said Tse, "we would have something like four hundred breweries." But that's not all. The fact that not all beer being consumed in Portland was local craft (though more than half of it was, he said) suggested to him there was room to grow, there and here.

The comparison had inspired Tse to invoke an analogy well suited to the province. He described the growth of the industry in Alberta as a line shaped like a hockey stick. For years, we just travelled along the blade. Now, as far as he is concerned, we've only just started our way up from the heel.

Alberta, of course, is not Portland. Not even Edmonton, an anomalous left-of-centre holdout in an overwhelmingly conservative province, is quite so quirky, achingly hip, or unabashedly admired. But a national comparison may bear Tse out, particularly if considering the Maritimes. In 2018, Nova Scotia, New Brunswick, and even tiny Prince Edward Island, for instance, were able to support between 7 and 8 breweries (of all types and sizes) per capita (that is, one hundred thousand adults of legal drinking age). At the same time, Alberta, a province with higher median and disposable incomes, had reached just 3.3 breweries per capita.

Tse's growth projections come with just one caveat. He does not see an Alberta of the future awash in the product of large beer companies. "In terms of breweries that are brewing twenty thousand hectolitres at a time and packaging and needing to sell thousands of cases of beer every month, I think that market is saturated," Tse told me. "How many Big Rocks can we have? One."

(It's worth noting that our conversation happened before Wild Rose was purchased by Sleeman, a move that, like most industry observers, Tse did not see coming. Could an influx of capital position the veteran of the Calgary brewing industry as a rival to the province's largest craft brewery?)

Instead, Tse foresaw little breweries popping up in neighbourhoods to fulfill a local need, as if alongside the resurgence of the independent baker, butcher, and barber. "There's lots of opportunity for smaller breweries that behave more like restaurants, where they're just selling to their local community. I think that's a very viable model for dozens and dozens of breweries."

While Tse's expectations seemed high, he was not alone in doubling down on more growth and lots of it. But like most observers and even brewery owners, he's no Pollyanna. He can see changes brewing. Some start-ups were not doing as well as they had hoped, he said, and others were looking to execute on exit strategies and sell while the demand for facilities and equipment remains high. But no one saw a great correction coming for beer in the province. The overarching belief was that it was not a bubble swelling, but a foundation being laid. Even if some breweries closed (in spring 2019, Wood Buffalo Brewing Co. of Fort McMurray did, making it the first of the boom to do so; Calgary's Red Bison Brewery, open for just a year and a half, followed in August; not long after relocating from Fort Saskatchewan to Edmonton, Two Sergeants Brewing closed that fall), peak beer was not visible on the horizon.

That may beg the question then of just how high could Tse's personal beer sampling tally go? More importantly, how much could it be boosted by Alberta beer? Judging by the strength and leadership of certain breweries—in building sound, bubble-proof businesses in addition to making great beer—Tse's notches may not yet reach very far up the hockey stick. The shape of his prediction may still also reflect the learning curve as much as the industry's potential.

But there are some participants who have already set the bar high enough to give themselves more than a head start.

BEST IN SHOW: TRENDSETTERS, RISK TAKERS, AND GAME CHANGERS

Alberta's can-do attitude sits somewhere between cliché and defining characteristic. This notion goes back decades, originating most obviously with the development of the oil sands. Sticky, tarry bitumen wasn't hard to find. It oozed thick and slow from the banks of the Athabasca River and the region's Aboriginal peoples are said to have used it to seal canoes. Exploiting it as the resource we know it as today was not so simple. There were no gushers like those that ushered in the southern US oil boom of the early 1900s. Nor was there anything resembling a Leduc Number 1, the rig that on February 13, 1947, sent a geyser of light crude fifteen metres into the air. The oil sands would never prove quite so eager.

To make bitumen into anything resembling the stuff that has become the foundation of the world's economy, it first had to be separated from the clay, sand, minerals, and water to which it was bound. As early as the 1920s, we found a way. At the Alberta Research Council, Dr. Karl Clark developed a hot-water extraction process that remains the basis of modern separation technology. With the funding and urging of government, it led to the development of an industry of gargantuan proportions, a vast industrial complex that represents an estimated $217 billion worth of capital expenditures. Like it or not, it is *the* symbol of innovation in the province, but the craft beer industry has its own Karl Clarks who have either tackled its stickiest problems or brought new thinking to galvanize and elevate it.

Still, just because a brewery makes craft beer, doesn't mean anyone will come. The word must be spread. For small businesses in general, the recommended spend on advertising is 5 to 10 percent of annual revenue. As that revenue won't likely be much at first, that probably seems like a lot. But to take short-cuts is to court failure, suggested Ben Keene, editorial director of BeerAdvocate, an online resource for all things beer, during a keynote at the 2019 Alberta Craft Brewing Convention in Edmonton.

"If we haven't heard of a brewery after it's open," he said, speaking as a beer geek tapped into the community, "then you're not doing a good job of marketing."

This seems like a low bar. The "we" Keene speaks of is the craft beer fan actively looking for something new to try. But attracting the fans may not be enough. Marketing can't just speak to the congregation—it also has to preach to the unconverted.

Big Rock led the way in doing this, years ago, when it launched the Eddies, the annual homegrown multimedia event the brewery once staged, beginning in 1993. The legend underlying it is that founder Ed McNally didn't like a price tag on an offer for TV ads, so he invited the community to do the work instead, and showcased their commercials as art. By the time the final instalment of the event ran in 2016, it had evolved into the Big Rock Eddies Short Film Festival (it now makes for fun YouTube bingeing). But the genius of it all was that McNally succeeded in soliciting what today is hailed as one of the most desirable and authentic forms of marketing: user-generated content (which was given credibility by the fact that McNally did not even require mention of the brewery's beer in the ads). The program effectively rallied a community around a product.

In a way, Village Brewery picked up where Big Rock left off, and also focused not on ads but on content. The approach had become more important than ever. These days, we're exposed to hundreds of messages (some industry analysts say thousands) that we really don't want to see or hear and perhaps have even become numb to. With that almost certainly in mind, Village Radio launched in 2013.

"Our original goal was to make beer and distribute it in Calgary, and also be involved in the arts and give back at least 10 percent of our net profits to the community," co-founder Larry Kerwin told me. Village Radio is an expression of that mandate. Hosted by veteran Calgary media personality Dave Kelly, the podcast features often fascinating, informative, and entertaining conversations about people doing good in the city, and has covered everything from arts to athletics to business to community development and much more. And it is almost never about beer; the Village brand is subordinated by any story that's being told. Nevertheless, listeners end up knowing the brewery because of it.

Of course, both the Big Rock and Village examples are products of deeper pockets than most startup breweries enjoy. Brewers operating on a shoestring budget need to be clever about the kind of message they send out into the community; they have to be smart about how they send it. One of the most intriguing strains of this kind of non-marketing requires little more investment than what might go into a batch of beer, and it's an example of the collaboration for which the craft industry is known throughout North America, Alberta included. In its most common form, the phenomenon manifests in a couple of brewers coming together to produce a beer baby, a hybridization of recipes and beer-making philosophies. The more interesting examples have involved more than beer. In summer 2018, Bent Stick Brewing applied the practice to independent music with Edmonton-based band the Wet Secrets. The idea: What if the brewery were to make a beer to pair with a set played by the band?

"[As] a scrappy independent band it made perfect sense to be partnered with a scrappy independent brewer," said Lyle Bell, lead singer and bass player for the dance-rock band.

The Wet Secrets, said Bell, "is this gaudy performance art, to some degree," with the gaudiness evident in the group's red-coated marching-band attire. In response, Bent Stick brewed up a red session ale. Bell and company provided the name, Hopstache, in honour of its 2008 track "Grow Your Own Fucking Moustache, Asshole."

"For brevity's sake it was reduced to 'Hopstache,'" said Bell, "which was probably wise for marketing."

Before the show, held at a downtown Edmonton music venue called the Common, a small number of ticket holders were treated to a ride on the city's High Level Bridge streetcar. During summer months, the Edmonton Radial Railway Society operates a short line between the south side and downtown, running an antique streetcar across the city's iconic, century-old bridge high above the North Saskatchewan River.

For a short while, they parked it in the middle and tapped a cask of Hopstache.

The effect of the non-marketing project on Bell himself may be the best illustration of its impact. Bell's not much of a drinker, though from touring with the Wet Secrets he was aware of the prevalence of craft beer in the US. Hopstache got him looking at the industry, and its role in the culture of his city, in a different light.

"It's cool that Edmonton can do that sort of thing. It's kind of cosmopolitan."

No doubt, some attendees of the show felt the same way. The experience was the thing that mattered, much like watching a film or listening to a podcast; it just so happened that a local brewery called Bent Stick helped make it happen. Unlike most of the hundreds of marketing messages encountered in the hours before that first sip of Hopstache, maybe this one, like the closing notes of the Wet Secrets' set that night, resonated.

Labelling is also part of the marketing rethink. Science says that, when it comes to food, we consume with the eyes first. The way food looks tells us, we assume, something of what it will taste like. If we're in a position to choose between one thing and another, this evaluation will factor into what's left behind.

Beer is the same but different. Since clear bottles permit light that can alter the flavour, no one gets a look at a lager or ale itself that might influence our decision to pick one or the other. Instead, the packaging is proxy for the product.

"How often have you bought something just because you liked the label on the bottle?" David Owens, chief operating officer of Edmonton's Sherbrooke Liquor, told me. "It happens all the time."

With the deluge of options available to consumers today, that's become more important than ever. "If nobody's tasted your product, you'd better make something pretty." More often than not, he added, choosing for looks is rewarded. "If [producers] are taking time to make a great product then they're taking time to make a great label. I've rarely had something that wasn't good that was in well-designed packaging."

Ribstone Creek Brewery is of similar mind. Opened in Edgerton, two and a half hours southeast of Edmonton, the brewery is among the geographically removed that have to work a little harder to get noticed. That's where Patrick Kerby comes in.

Kerby wasn't looking for a job with Ribstone in 2012, not long after the brewery started up, when he tagged them in a tweet while enjoying one of their beers.

"I took a picture of the glass on the table, put some filters on it, and posted it," the thirty-three-year-old independent graphic designer told me as we sipped coffees in a French-inspired café near his home in east Edmonton. The brewery contacted him to ask if they could use the shot on its website. He said they shouldn't—they should have something better. His offer of pro shots (his partner, Bri Vos, is the real photographer in the family) led to eventually being put on retainer as the brewery's designer.

I had initially reached out to Kerby for similar reasons, in that I noticed his work as I browsed a liquor store shelf. It was the packaging for Old Man Winter Porter that caught my eye, an evocative personification of our most miserable season and almost classically artistic. It was an approach to packaging that, at the time, I'd not seen in Alberta craft beer. And it was the reason that I bought that six pack. With beer, the aesthetics are as thin as the average aluminum can, but they're hardly superficial. Kerby thinks of himself more as a designer than an artist. He has a client to satisfy, which requires that he be able to, at some level, satisfy the customer.

"My occupation depends on a consumer relating the care that goes into the marketing materials [to] the product as well."

The relationship between craft beer drinkers and craft beer brewers, as he understands it, is not merely transactional. It's more personal. If small-batch brewing is an expression of identity, so too is choosing to buy the product of one or another.

"It's not just like, 'Here's some beer. Aren't you just glad that there's some beer available?'" said Kerby. "It's the ingredients, the care that they're putting into it, and that's what they're selling."

Kerby and Ribstone are not alone in their efforts to connect with buyers in a way that might apply just as much to an artist's attempts to appeal to gallery-goers. Calgary's Dandy Brewing Company sponsors an artist residency at the brewery that has clearly influenced its approach to product design. There's no direct connection between the two, as artists are not required to produce anything for the brand, but the company's labels bear a sense of whimsy, style, and even humour that embody the notion of the dandy, eccentric and confident. The company is also a good example that pretty need not be the objective of packaging. The unexpected will sometimes do just as well, as this, too, tends to be an excellent quality in a beer. Case in point: That is, indeed, a sketch of a moustachioed octopus on the brewery's best-selling beer, Dandy in the Underworld, a sweet oyster stout.

Blindman also keeps several artists and designers employed in providing content for their labels. One particularly striking result was the art for its limited-run 24-2 Stock Ale, the beer made with barley the brewers helped grow, an effort that was in turn helped by twelve stock horses. Amie Chaotakoongite, co-owner of Avenir Creative in Edmonton, highlighted the animals' roles by capturing them mid-stride in a dozen unique, minimalist line drawings (along with the horses' names). Each was affixed to separate cans randomly packaged into four-packs, as if the brewery were issuing a challenge to collect them all.

It was tempting. Like the beers Kerby designs, and just as Owens suggested, the beer—a hearty brown brew with suggestions of dark fruit and rustic bread—could be judged by its cover.

The 24-2 Stock Ale is, in many ways, a kind of symbol of the future potential of Alberta craft beer. That might be contentious, to highlight one beer and one brewer given that brewing talent is unquestionably now a defining characteristic of the industry. The 2019 Alberta Beer Awards proved as much. Alley Kat showed that an old cat could still learn new tricks, and was named brewery of the year. Lethbridge's Coulee Brew Co. came in second, largely for exceptionally well-made beers that tend to cleave closely to familiar styles, which could also be said for award-winning breweries such as Apex Predator, Fahr, Troubled Monk, Tool Shed, Wild Rose, and more, though none of these have shown any fear of experimentation. Then there are up-and-comers known and lauded for consistently pushing limits, including Annex Ale Project, Snake Lake, Town Square (the bronze winner for brewery of the year), Grain Bin, and others. (The number of award winners at the 2019 awards who'd opened their doors the year previous, was, frankly, astonishing.) Many readers will feel that list is marred by glaring omissions. Which is a good sign. But since this book is descriptive rather than exhaustive, we'll draw toward a conclusion by looking more closely at two breweries that to me represent the ambition and potential of Alberta craft beer in the post-2013 era.

That 24-2 Stock Ale is actually but one example of the innovation for which Blindman has become known, and of the lengths to which they will go to produce unprecedented beers. Sure, they grew, harvested, and malted barley for their own beer, but there are even more extreme examples of what these guys will do for a brew. They're called foeders: massive oak barrels that, in a way, wrap up everything the brewery strives for. That includes nerding out as brewers, telling good stories about why and how they brew, growing as entrepreneurs, and, of course, making amazing beer. But that's getting a little ahead of the story.

Blindman has roots in both Lacombe and Edmonton. In the capital, immediately post-deregulation, Kirk Zembal had an eye on the scene, convinced it was heading in the direction he'd seen during beer trips to the US with his brother-in-law, Matt Willerton, who'd worked in operations at Alley Kat. It wasn't just the legislative changes that had him excited. Zembal went so far as to corroborate his hunch with keyword searches on Google, finding "craft beer" coming up with increasing frequency in the region. It got him and Matt to thinking—along with friend Shane Groendahl of Edmonton Beer Geeks Anonymous—that the time was right. But as they looked into zoning challenges and costs, it occurred to them that the place was not.

Edmonton, said Zembal, "was a no-go."

In the meantime, Dave Vander Plaat and Hans Doef were independently hatching a plan in Lacombe to start making the most of being in the midst of Alberta's richest barley-producing region.

"Basically, in the beer world you just hear things," said Zembal. "I heard these guys were going to open up a brewery in Lacombe, so I thought maybe I should reach out to them." Soon enough, a partnership was forged, "as all good decisions are made—over beer."

With the City of Lacombe proving amenable to the idea, and not bogged down by big-city bureaucracy, Blindman opened in an industrial park not far from downtown, and sold its first beer in October 2015. Ever since, they've managed to keep to a shared philosophy of beer making that to this day shapes the brewery's ever-growing output.

"We've always done what we wanted to do," said Zembal. "There are beers we like to drink that we couldn't get in Alberta at the time. There weren't the IPAs we like to drink, or golden ales with a good hop punch. We really like dry beers, English-style beers, stuff we couldn't get. So, we just decided to make them. We like to brew what we like to drink."

Zembal knows that might sound somewhat exclusive. It's not meant to be. Alberta, he said, is a "beer-flavoured beer market." That is, he feels it's still dominated by light lagers made by multinational companies that aren't likely to stray far from business plans—and recipes—that have served them well for decades. For him and his partners at Blindman, however, craft beer offered a kind of awakening, an experience that Zembal feels the brewery could inspire in other drinkers.

"None of us came out of the womb drinking craft beer. We've all had that aha moment. And we want to be that brew for people."

When the possibility of foeders came up, they seemed like a new way to "be that brew," in this case, what Zembal would describe as "funky beer." Foeders also seemed incredibly idealistic. For one thing, they aren't cheap. For another, Blindman was only a year old when they decided to push ahead with buying them and having them shipped over from France. Money was in short supply, and enough had already been borrowed that going back to the bank was not an option. In fact, it's fun to pause a moment to imagine what might have transpired had they paid the loan officer another visit.

"Kirk, good to see you again! How are things at the brewery?"

"Awesome. Growing like crazy. In fact, we're having a bit of trouble keeping up with demand."

"Wow, and just a year into it. Fantastic. So, what can we do for you today?"

"We've got a new project coming up. We're pretty excited about it."

"Excellent. Tell me more."

"Have you ever heard of a foeder?"

"Excuse me?"

"A huge barrel used for fermenting beer."

"How huge?"

"About three thousand litres."

"Whoa."

"Yeah. They're used a lot in Europe for things like wine and cognac but they're catching on in North America. We'd like to bring a couple of second-hand ones to Lacombe. We think we could use them to make some pretty amazing stuff."

"Sounds interesting. You'll need some support for that, I gather? What are the challenges? Lay it out for me."

"OK. First there's the cost of the foeders themselves. We want two of them. We have to get them shipped over, and that isn't cheap. Then once they get here, they'll need some repairs. These things are, like, forty years old. So, we'll have to find a cooper, a guy who fixes barrels. Probably have to fly him in. After that, we brew up a few thousand litres of wort and put it in the foeders and wait for whatever kind of microbes the previous owners left in the wood to start working their magic. It takes a while, months, years even. Then we can draw off some beer and start selling. You can't sell the whole batch at once or else the wood dries out. They tie up a lot of floor space because the product just sits there for ages. But we're absolutely certain that these things are going to make some great beer—and there are very few breweries in Alberta even thinking about this right now. We want to get going on this as soon as we can. Oh . . . and they're not exactly money makers, to be honest."

"OK . . . great. Hey, have you ever considered crowd-funding?"

In reality, Zembal did not need a banker to suggest the idea of crowd-sourcing funds for the barrels directly from drinkers themselves. They felt it was the way to go from the start and so would accomplish three things. First, it would raise money that was required. Second, it would let them indulge their passion for the weird and wonderful in the world of beer. Third, it would allow them to make a statement to Alberta drinkers that weird and wonderful was Blindman's thing. It essentially reinforced everything about their brewing philosophy.

"It's always good when we get to talk about what we're excited to do," said Zembal. "I think people respond to that."

That philosophy was implicit in the inner workings of a foeder, which, on the surface, is very simple. Wort, beer's sugary precursor, is brewed and transferred to barrels where it's inoculated with a mix of yeast and bacteria, which commingle with the mix of yeast and bacteria that's likely already present in tiny grooves and crevices in the oak of the foeders. After that, the beer ages and a collusion of microbiological and biochemical processes over several months leads to a unique flavour profile. The end results are often unpredictable and "funky."

"We love making those kinds of beers," Zembal told me. "There are people who didn't get it. Most people hadn't heard of foeders before." (Or, for that matter knew how to pronounce the word. I didn't. It's *food-er*, if you're wondering.)

Through social media, newsletters, and updates on their Indiegogo page, Zembal shared the progress of acquiring the barrels, including having them repaired by a Frenchman named Jean-Pierre (the two foeders now bear his name: Jean and Pierre) before they left France and again on-site by a cooper brought in from New England. He then covered the development of the beer. It was a story of Alberta beer that had never been told before, as there had been no reason to tell it.

"Stories have the power to move us," BeerAdvocate editorial director Ben Keene said during his keynote at the 2019 Alberta Craft Brewing Convention. The key lies in "compelling [an audience] with a message to want to know more, to want to learn more."

By the time the campaign to raise funds for the foeders wrapped up in February 2017, 186 people had contributed, including me, who tipped in a measly $50 in exchange for the promise (or perks, as they're called in online fundraising efforts) of a couple pieces of stemware and a few bottles of the first beers eventually drawn from the barrels. They'd asked for $20,000 and came close, raising $17,055. It was an impressive run, given that Blindman had a relatively short track record, being just a year old, and, even more to the point, that they wanted to make a beer most Albertans knew nothing about and might not even like. Neither proved to be a negative in the end. Through the beers it has made so far, including a kettle-soured series previously unseen among Alberta breweries, it has apparently proven itself. The first foeder beer, simply called "Pierre" after the barrel in which it was fermented, was released in February 2018, roughly seven months after the foeders were filled. I am qualified only to say that I enjoyed it immensely. A crowd-sourced review at beeradvocate.com described it as "funky... earthy... musty... worth checking out, as a new day dawns in this burgeoning Alberta craft beer scene."

Zembal would likely chuckle at that. "We won't pretend that we are changing the world by embarking on this path," he'd written on the Indiegogo page. What they were trying to do was keep their brewery evolving. With the foeders, he'd added in his fundraising appeal, "we will continue to find new ways to keep pushing the boundaries of the beer we make." And with it, the boundaries of what defines beer in Alberta.

THE NAME THE Monolith is obviously, even eerily, appropriate for Greg Zeschuk's somewhat intimidating-looking Edmonton brewery. Or at least it's obvious if you consider those dark, seamless slabs that punctuate Stanley Kubrick's 1968 film, *2001: A Space Odyssey*. The building is tucked away on a quiet back street not far from Whyte Avenue, Edmonton's prime party district, right across the road from what used to be Amber's Brewing Co., one of the city's early and now-defunct craft brewers. When I first saw The Monolith from the curbside in spring 2019, its taproom still under construction but the rest of it functioning, it was not clear what the place was. The blocky building's dark facade bore no name, etched instead with the logo of its parent, Blind Enthusiasm, which Zeschuk describes as a "market brewery." Situated a few blocks east, it comprises a butcher shop, coffee kiosk, a small brewery, and a restaurant called Biera. At the centre of the logo that graces both is an opening eye that lends it a secret-society feel. Or perhaps that of an awakening. All that was missing during that first visit to The Monolith was the horns and timpani drums of Richard Strauss's *Also Sprach Zarathustra* with Kubrick's camera catching the sun rising behind it.

Maybe the name is meant in jest, an extension of Zeschuk's smart and subtle sense of humour. But follow the reference back to the Arthur C. Clarke science-fiction novel on which the film was based and the gravity deepens. The appearances of the monoliths around the solar system were harbingers of societal and cultural shifts. Writing about the movie on the eve of its fiftieth anniversary in 2017, University of Texas professor and art critic Barry Vacker re-examined a scene in which apes encounter one of the monoliths in the desert: "When the apes first see the monolith, they are experiencing an all-too-human moment of the sublime—the simultaneous feelings of awe, wonder, and terror when gazing upon something majestic and mysterious that seems to overwhelm their reason and sensory perceptions."

OK, I can't help myself. The ape reference aside, it's too tempting not to draw a parallel with the advent of craft beer in Alberta, at least in that sense of awe, wonder, majesty, mystery, an overwhelming of reason and perception. Isn't that what we all hope for every time we crack open a new craft beer? And now we have The Monolith, which has dropped down into the province as an agent of change. As a brewery focused entirely on barrel-fermented beers inspired by a Belgian style known as lambic, Zeschuk's project has no rival in the province, if it indeed does in Canada.

The irony, of course, is that it is not a sleek and modern new way of doing things (even if the building looks it). Zeschuk and head brewer Doug Checknita are making beer by a method developed in Europe centuries ago. Replicating it required a level of dedication that most brewers would describe as a luxury if not an impossibility. The Monolith practice involves an elaborate, traditional process of boiling water drawn off of the mash, the beer-making step in which crushed grains are steeped so that their starches are converted to sugar. The boiled portion is then re-added to that mash.

"Mashing a normal beer takes an hour, hour and a half," Zeschuk told me in his impeccable, purpose-built facility. "This takes three hours."

But The Monolith team sees this as the best way to produce wort that approaches the quality and flavour of the stuff made for centuries in Belgium. It's later flavoured with dried whole-leaf hops rather than pellets, and boiled for four hours, rather than the usual one hour that suffices for most brews. The old-timey innovation does not stop there. All this effort is but the precursor for a time-consuming step that pushes back potential profits on that beer into what, for most breweries in Alberta, would be the much too distant future.

"It's interesting because making the wort is not the creative part of the process," said Zeschuk. "It's raw substrate that you put into barrels... and then you kind of cross your fingers."

After it's cooled, the fresh wort is transferred either to a holding tank, where a mix of yeast and bacteria are added or, depending on the time of year, it's pumped directly to a coolship in an enclosed room on the top floor of the brewery. Custom-made, that circular stainless-steel tub is big enough to hold twenty-five hundred litres. Essentially, it's a giant petri dish.

In mid-spring, once the snow melts, or in the fall before freeze-up, environmental conditions are best for supporting a robust population of naturally occurring yeast and bacteria that float about in the Edmonton air (which, incidentally, is every bit as good for initiating spontaneous fermentation as that of Brussels). These ride a breeze through louvres that open into the room that contains the coolship. That's what Zeschuk and his brewers hope, anyway. The next day, before any evidence of microbial activity is seen, the presumably inoculated wort is pumped to the holding tank. Twenty-four hours later, it goes into oak barrels sourced from Europe and the US, which are then sealed for as long as three years of fermentation. They're stored in windowless concrete rooms that are kept at 12 to 15°C and humidified by overhead atomizers that send water down in a state somewhere between a mist and a fog. Though the walls were nearly pristine at the time of my visit, Zeschuk hoped that one day they would look like those of an old-world cellar, discoloured by a layer of mould and bacteria that would potentially, and somewhat mysteriously, benefit the process. Ultimately, barrels would be blended for bottling according to the brewer's tastes, a process that Checknita is familiar with from his time working at Cantillon, in Brussels, one of the most highly respected breweries in Europe, if not the world. Access to the barrel rooms is, by the way, restricted by a card-recognition system.

Regardless of the name, The Monolith was not actually created to usher in a shift in Alberta's beer culture, or any culture for that matter. "Why on earth would we build it?" Zeschuck said of this wondrously high-tech homage to what was previously a bygone era of brewing. "Because, well, why not?"

That's an answer one might expect of him. Now in his early fifties, he self-financed Blind Enthusiasm, which opened in summer 2016, and The Monolith largely on the success of a previous venture. After spending a few years practising as a medical doctor, he co-founded a video-game company in Edmonton in 1995 called BioWare, which was sold in 2005 to a private equity company based in Silicon Valley for an undisclosed sum. In 2007, it was sold again, this time to Electronic Arts, for US$860 million. At the time of this sale, the original owners of BioWare, said Zeschuk, still had some ownership stake in the company. Whatever he may have earned from the transaction has mobilized a passion for craft beer into which he took a deep dive during time spent in the US, Austin in particular, where he continued working in the gaming industry after the acquisition. There, he also launched a web series called *The Beer Diaries*, a craft-beer-fuelled travel show that highlighted the world's best breweries and most interesting beer scenes.

"What have you invested in all this?" I asked Zeschuk at the end of our tour of The Monolith.

"I'm not answering that one," he answered with an amiable laugh. "Lots."

Zeschuk hoped to have the first 650- or 750-millilitre bottles released by fall 2019 (he did, by mid-December), and expected them to retail for $25 to $30 each. When I winced at the price, he shrugged.

"That's the point," he said. The Monolith is a "boutique" operation, Zeschuk pointed out. "Low production, high demand. You charge more. It's like the wine industry. It's potentially very profitable if you succeed. I mean, it's got to be good. If it's not good, then the whole thing is a disaster."

Edmonton-based beer writer Jason Foster believes that Zeschuk's beer will be a hit with the province's craft beer drinkers. But he wonders if its impact might be more pronounced outside the province. Zeschuk has plans to export once he's producing enough, potentially making The Monolith "a bit of a game changer in terms of how the world sees Alberta," said Foster. If it attracted an international beer critic to see what all the fuss is about, he or she might not head home immediately after, he added. "That's nothing but good because while he's here... lo and behold, [he'll] discover the range of diversity that is Alberta beer."

Arthur C. Clarke once wrote, "The only way to discover the limits of the possible is to go beyond them into the impossible." His monoliths were, in a way, invitations to make that leap, and to recalibrate our understanding of those limits. What is craft beer in Alberta? It should be something that no one can quite put their finger on; it should be something that did not previously exist. Most of all, in the right hands, it should be something capable of changing how we see what's around us, whether it's a drink, an industry, a city, a province.

If anything, that's how Zeschuk views his responsibility as one of Alberta's leaders in craft beer.

"If you can do it, then do it," he said. "If you have the ability to do something really cool and really fun, especially in your hometown, do it."

CONCLUSION
Riding the wave of Alberta craft beer

WHILE THE DAM has broken, and, other than to strip entrepreneurs of their businesses and brewing licences, there's no conceivable way to seal up the industry again, Alberta craft brewers are not quite yet riding a river in unimpeded flow. Just as in any new industry, rough waters arise here and there. Overall growth, as Don Tse suggested, will almost certainly continue, but some breweries will go under, simply because, in one way or another, they're not respecting the current. When I spoke with Greg Zeschuk in spring 2019, he echoed the oft-expressed belief that Alberta was nowhere near craft beer saturation at 110-plus breweries. But he added a caveat.

"I don't think we're reaching saturation. We're reaching saturation of mediocrity. That's where the danger is, if you're mediocre." There's only so much shelf space at the liquor store, he points out, and so many taps at the local bar, and nobody's bottom line can accommodate "good enough" for very long. They're looking for the "best of the best," he said.

Former Alberta Small Brewers Association executive director Terry Rock (who succeeded Zeschuk in the role in 2017) noted the danger in any new brewery making the assumption that, simply by virtue of making beer in small batches, they'd succeed in endearing themselves to the community. It's not that simple, Rock said. "You have to have a different take."

One reason for any sense of sameness in the industry today is that it's still young, said beer writer and entrepreneur Don Tse. "We're still trying to mimic. You see that in the beers that most of these breweries are making. You see IPAs and ambers and blondes. A New England IPA becomes a hot thing so everyone brews a New England IPA, and then a brut IPA becomes a hot thing so everyone brews a brut IPA."

What does a standout brewery look like, then? We know that very few of them will look anything like The Monolith. But Tse felt they could follow the example of the likes of Dandy in Calgary or Lacombe's Blindman. What they can't do, however, is stand out for the wrong reasons, which include using sex to sell a product, a practice that is being questioned even by multinational companies that leaned on it for decades (some still do, of course).

"The world has moved on from lads telling jokes on a Saturday and high-volume consumption," then SABMiller CEO Alan Clark said in 2015, when global beer sales were stagnant. "Beer is now drunk by women and men together."

Surely, the Alberta craft community is so enlightened, you might say. Not all of it, unfortunately. Packaging for the core offerings at Fat Unicorn Brewery, located near Plamondon, features drawings of women to match beers named Naughty Amber and Dirty Blonde. Half Hitch Brewing, with locations in Cochrane and Canmore, veers unnecessarily into questionable territory, as well. At the time of writing, the can for its red ale represented part one of a story of a young couple, the woman struggling to cover herself after being discovered in the barn by a shotgun-bearing rancher, presumably the father. The cartoons may be an attempt at humour, but why make a joke that involves a woman in distress?

Liquor retail consultant Shelly Hall, who points out that women make up a craft beer target demographic with enormous growth potential, was concerned that such labels speak to something bigger. "While these labels don't offend me in any way, they do make me question the level of social consciousness of the breweries that use labelling which could be construed as sexist," she said.

For her, it comes back to that moment in the local beer cooler, when the roving eye of a customer finally comes to rest on something that pleases her or him. "As consumers of all genders have more and more options to select from, the less socially conscious breweries"—even if they're only mistaken as such because of a joke that's backfired—"will change or they will fail."

As the industry grows, and liquor store shelves become ever more crowded, brewers may also court failure by not thinking beyond them—that is, beyond Alberta. Of course, simply crossing provincial boundaries may not be enough. Nor is it particularly easy. The cruel joke that many of the province's brewers ruefully tell is that it's easier to send beer out of Canada than to try to sell it across the country.

"We've sold beer in South Korea," Alley Kat Brewing co-founder Neil Herbst once said, "but we can't sell beer in Ontario or BC, or for that matter pretty much any other province."

The Alberta market is the most open in Canada, shaped by whatever its private liquor stores and importers choose to bring in. As of March 31, 2018, there were 4,946 beer products available at local liquor stores. By comparison, Ontario, Canada's largest liquor market, saw 1,246 domestic products on store shelves in 2018 and just 383 imports.

The difficult process of getting listed by a government-run liquor retailer in another province was one of the factors that motivated Tse to become an exporter—to other countries. In winter 2018, he co-founded Far Out Exporters to help Alberta breweries send products abroad, managing matters such as logistics, risk management, and foreign exchange. By spring 2019, he said he had "about a dozen and a half clients." The company has found that certain global markets, particularly developing nations of Southeast Asia and even parts of Eastern Europe, have a taste for Western imports of all kinds, including beer. And Alberta's will do.

The benefits of moving more volume aren't just increased revenue, Tse believes. He acknowledged his bias, but his rationale was sound. It focused on improving the economies of scale through making more beer for multiple markets, and on churning through ingredients more quickly, while they're fresh, leading to better beer locally.

"I think [exporting] is an opportunity for them to up their game," said Tse.

It's not just the breweries that need to up their game. It's governments, too. There's a mystery in the Alberta craft beer industry that, if not quite unsolved, is going unaddressed. It's embodied by a frustrating tale of two cities—that is, the reason for the difference in the number of craft breweries in Calgary and Edmonton. At the time of writing, there were approximately twice as many operations in Calgary than Edmonton. The cities have similar populations, and likely a similar portion of those are craft beer enthusiasts. They also both have veterans who helped ready the market for when the government removed the barrier to entry in 2013. So, what gives?

Jason van Rassel would not hesitate to take anyone to task for suggesting that it has something to do with differences in creativity or entrepreneurial spirit between the two jurisdictions. He's lived in both and loves both. Some suggest it has something to do with the recession freeing up capital from the more natural resource–intensive Calgary economy. Maybe. But he refuses to accept the idea that Edmontonians might not be willing to put skin in the game.

"Not to take away from Calgary's entrepreneurial spirit at all, but the simple fact of the matter is that Calgary does not have a monopoly on entrepreneurship," van Rassel said. As proof, he pointed to The Monolith, a risk despite being representative of serious thought-leadership in craft beer in Alberta. "If that isn't entrepreneurial spirit, I don't know what is."

A hint about the causes of the discrepancy came from one of my conversations with Kirk Zembal, co-founder of Blindman Brewing. Without disparaging city council or administration, he said of startup governance in Edmonton that some things are just easier in smaller centres, where there's easier access to city council and managers. Zembal wouldn't say more than that.

But John Toman would.

In May 2019, he and four other partners opened Odd Company Brewing in Edmonton's Oliver neighbourhood, not far from downtown. It's a highly experimental brewery where it's not uncommon to find things like a raw pale ale or a white stout on the menu. And it's the product of what Toman describes as relentless pressure on city administrators. A building permit that should have taken a month, he said, instead took four, pushing back the brewery's opening date.

"You kind of have to be an asshole and be persistent," said Toman, who at one point was sending around four emails a day to get responses. "You have to annoy them into dealing with you."

Adam Corsaut had a similar experience with opening Analog Brewing, that video-game-themed craft brewery on the city's south side, in July 2018. "You would get told different things on different days based on who you spoke with," he recalled. That applied to inspections as well. In one instance, he recalled, a city inspector for a water line paused to fail a previously approved inspection for a gas line, requiring a new inspection of the line (including another fee).

"The federal and provincial governments were very easy to work with in comparison with the city government," said Corsaut. "When I contacted the city to try to straighten things out, I always got shuffled around between departments and never really got an answer. And if I did get an answer, I'd get somebody else a week later and it would be contradicted."

Part of the problem, suggested Toman, was a feeling that certain staff members lacked confidence in the information or advice they were sharing.

"It seems like a lot of people who work there, their main goal is not to get in trouble," he said. "Literally their goal is not to get shit on by one of their bosses. That's my estimation of how the city operates."

That culture has been extensively documented. In 2018, an employee engagement survey indicated that nearly a quarter of 8,732 respondents reported being harassed at work. That was up from 19 percent in 2016. The City of Edmonton continues to fight culture and effectiveness battles on numerous fronts.

Toman did say that since opening, the relationship with the City has been fine. Corsaut said that the experience probably made him and his business partner stronger, more resilient. He believes that the mayor and council have had the interests of small businesses like his at heart. In 2017, they voted to amend zoning bylaws to relax certain location restrictions. Nevertheless, said Corsaut, the administration at "the City of Edmonton caused our greatest delays and greatest confusions."

He was not hopeful that the underlying causes, which he believes contribute to the stark disparity in the number of breweries in Calgary versus Edmonton, would be soon resolved. The upside, as he saw it, was grim. "In the future, when we need to do renovations and get permits, we at least know to expect delays, confusion, misunderstanding, and lack of communication."

And as much as Corsaut may appreciate the province's support of small brewers, particularly through what he saw as a helping hand in the AGLC, work also remains to be done by that level of government. When it comes to beer, the face of that government was Joe Ceci. There is no shortage of photo-ops of the former New Democratic Party minister of finance, and now MLA for the riding of Calgary-Buffalo, raising a pint at various craft breweries in Alberta. "Like I like to say, beer is good," he told me not long before Jason Kenney replaced Rachel Notley (a dedicated fan of IPAs, Ceci said) as premier. He meant that in an economic sense as much as anything else. The NDP government was not responsible for the changes at the end of 2013, but it eagerly picked up where its Progressive Conservative predecessors left off.

"We've worked to modernize Alberta's liquor laws and liquor manufacturing," Ceci said. "It's part of our government's plan to diversify the economy outside of oil and gas generally."

This was not necessarily a plan guided by hard data. Ceci did not know, for instance, the industry's true potential for job creation, as was suggested by its success in other jurisdictions throughout North America, including Ontario, where in 2015 it was responsible for some seventy-five hundred jobs. But his government knew enough to know that craft beer was "an area that was underappreciated," as was its underlying supply chain, right from "grain to glass."

"We want to see Alberta grow from having the best wheat and barley in the world, and it makes sense to turn that into great craft beer," said Ceci.

To that end, the government changed a lot of rules. It made possible (that is, legal) taprooms as the community-oriented craft beer destinations we know today in Alberta. In 2017, it let small brewers sell at certain farmers' markets.

Not all of the government's attempts at support, however, worked out quite so well. In 2015, it changed the mark-up system, the tax imposed on alcohol production, so that it favoured brewers in Alberta, British Columbia, and Saskatchewan, the three members of the New West Partnership Trade Agreement. Beer producers within the region would pay less tax than those from outside. When that program ended in 2016, Alberta reverted to a previous, higher mark-up to which all breweries selling beer in the province, local or not, would be subjected. The difference was that the province's small brewers were simultaneously made eligible for grants that matched the difference between the higher 2016 mark-ups and those of 2015. The effect was that an Alberta small brewer could potentially sell at lower prices than out-of-province competitors.

This brought on a trade complaint from a Calgary-based beer importer. Court challenges came from

Ontario's Steam Whistle Brewing and Saskatchewan's Great Western Brewing Company. The brewers (neither of whom responded to an emailed request for comment) argued that the changes violated sections of the Constitution Act. They wanted restitution. On June 19, 2018, Honourable Madam Justice Gillian D. Marriott gave it to them, finding that, first, "the essence of the 2015 Mark-up was to create a trade barrier related to a provincial boundary." As for the change that followed, it wasn't the grant program that was the problem. It was the way that was coupled to a tax. "The 2016 Mark-up and the grant program cannot be considered in isolation from one another." In fact, she added, "The Minister [Ceci] specifically acknowledged that the 2016 Mark-up and the grant work 'in concert.'" This, too, she ruled, was unfairly advantageous. Steam Whistle was awarded $163,964.98, Great Western $1,938,660.06. Both amounts were based on calculations of mark-ups that were ruled to have been unfairly paid.

After the Alberta government was told that it couldn't support its domestic industry at the expense of others, it realized it had to take a different tack. Ceci would not comment on the case during our conversation but, after Justice Marriott's ruling, he vowed that his government would not back down.

"I want small brewers and liquor manufacturers in this province to know that we'll continue to have their backs," he said. "We believe that diversification is critical and they're part of that whole initiative to see Alberta get off the oil and gas roller coaster."

With that, the provincial government launched a complaint of its own. Under the Canadian Free Trade Agreement, it alleged that the Ontario government's liquor policies discriminate against Alberta's small brewers. The Liquor Control Board of Ontario listed around twenty Alberta liquor products at the time.

The challenge is a considerable undertaking, one that will likely take years. The NDP, it turns out, didn't have years. Soon after, I asked Mike McNeil, executive director of the Alberta Small Brewers Association, about his hopes and expectations of working with the then-new United Conservative Party government. High among McNeil's priorities in his role, which he took on in spring 2019, is to break into the Ontario market.

"Certainly, our talks are preliminary," said McNeil. "I've had discussions with political staff. I sense that they are still getting up to speed and learning their files. They haven't taken a concrete position on issues besides interprovincial trade and reducing provincial trade barriers."

Travis Toews, who now has Ceci's old job as finance minister, responded to CBC News to say he wants to learn about these challenges. Toews did not respond to my emailed query about how Alberta hopes to deal with the trade imbalance. Help will be required, just as it was in the past. Mind you, it ought to be constitutional this time around. One Alberta industry insider I spoke with (off the record) agreed with Justice Marriott's ruling, on the basis that the mark-up regimes undermined the free market and may have allowed for the creation of businesses—that is, breweries—that were inefficient, and therefore possibly unsustainable.

Beer writer Jason Foster didn't see the mark-ups and grants in quite the same way. The industry needed government support then just as it will likely continue to need it in the future.

"Recognizing that it was completely trade-noncompliant, it addressed the big inequity," said Foster. "It actually did what every other province did for its breweries decades ago and it created a little bit of a safe space, dampening the competition a little bit. Some breweries pulled out. Some breweries who are relying on price point could no longer rely on price. It just kind of gave a little elbow room. And that's what the [local] industry needed. And I think that's what the policy did."

The results, in his view, speak for themselves. Today, Alberta has more breweries per capita than Ontario and Quebec, and nearly as many as British Columbia. "There's just no turning that clock back. All [the industry] needed was a couple of years of some space."

That those couple of years cost a couple million dollars does not concern Foster. "Speaking as a long-time advocate of the craft beer industry, I think it was totally worth it. Every penny."

MEANWHILE, BACK IN THE BREWERIES OF CALGARY

On that fateful and wonderful evening exploring Calgary's craft breweries, our ride-share dropped us off at Establishment Brewing Company around 8:45 PM. At this point in the tour, we'd have been highly likely to endorse Foster's sentiment and say that what the government spent on craft breweries was indeed worth every penny. But that didn't mean we were incapable of arriving at certain conclusions of our own regarding our experience among the light industrial streets of southern Calgary. I was struck, for instance, by the potential of the district as the true tourist destination Calgary wants it to be, as in, "Hey, this thing might just be crazy enough to work." In the building *next door* to Establishment was another brewery, Annex Ale Project. We'd be able to walk from one brewery to the other and it would take less than one minute. Nothing in my life in Alberta before this had ever been so simple.

The proximity, it turned out, was a blessing, as we were in no mood to hurry, having found that Establishment was a gem unto itself. It was a clean, softly lit place with plenty of seating that was mostly unoccupied despite it being the middle of a pleasant Friday night. Sloan on the sound system—on vinyl, no less—broke what would otherwise be near silence. Perhaps because we were three middle-agers, and because

Sloan occupied much of the soundtracks of our youth, we felt not lonely but welcome.

I ordered tiny glasses of pilsner, New England IPA, and a "summer ale" that smelled pungently of elderflower and balanced its light malt with fruit and floral undertones. I finished it and the pilsner and passed the remainder of the NEIPA to Colin. He tipped it back with aplomb. We shared a small Mason jar of mixed nuts, purchased at the bar for $4, and I was uncommonly delighted at the sight of a macadamia. Maybe this, too, was owing to Sloan, which had landed on the 1996 hit "The Good in Everyone." Guy, whose vegan meal at Dandy was tasty but not fortifying, carefully selected a cashew from the jar.

The nuts eaten and the glasses emptied, we made the short trek to Annex. "This is how it should be," I said, about thirty seconds later as we crossed from one parking lot to the next.

"Yeah, man," said Colin.

The reason that it was how it should be was not so much that it was all so convenient, even if that was an excellent feature of our taproom tour. It was that it captured that sense of craft beer as an experience, highlighting its diversity and the community it can foster. As we shared a long wooden table with a few other patrons at Annex, I sat back with samples of sour wit and rye ESB, styles I'd never tried before, and we chatted with strangers. Looking back later, it reminded me of something I'd once read by British travel writer Pico Iyer. "For me," Iyer said in a lecture he delivered at the Smithsonian Institution, "the first great joy of travelling is simply the luxury of leaving all my beliefs and certainties at home, and seeing everything I thought I knew in a different light, and from a crooked angle."

To sit and enjoy two unfamiliar beers at a communal table with people you've just met is to become vulnerable to the unknown. Leave your beliefs and certainties at the door, for inside you may find new ones. What I didn't know about Annex was, in part, what I didn't know about Calgary, which is, in turn, what I didn't know about Alberta. Here among these breweries, the twilight of the mid-spring night illustrated the fact that the province I call home remains, to me, largely uncharted territory, and that I am not only susceptible to its charms, but capable of being somehow changed by it, and—why not, at this point—perhaps even improved. Because of craft beer. That is a good thing.

But Annex was not the end of the experience. We checked the time, finished our drinks, and called for the ride-share, worried that the ball was coming to an end at Village without us. We had a concert to catch.

At about 10:30, we arrived at the brewery just in time to catch the last few songs of the band's set. Copperhead is a Calgary quintet fronted by keyboardist and vocalist Liz Stevens. They delivered a languid and dreamy pop set with vaguely sinister undertones attributable, I thought as I listened, to the band's baritone sax player. I went to the bar and order three glasses of the night's cask. The bartender gave me a worried look before tipping the nearly empty vessel up to squeeze out three small glasses.

This was OK, because the cask, which Village concocted based on instructions from the band, was not to our taste. A couple of weeks before the show, Copperhead came in and suggested it be three-quarters blonde ale, one-quarter dark, and flavoured with hibiscus tea, vanilla, and blueberries. Oh, and "skin flakes (all members)," said the sign by the cask. As one of the members of the band told me after the show, "It's a sweaty brew!"

"It's like drinking the bar mat," said Colin. He did not tip back his glass, and instead set it on the table, nearly full. But that's casks. Some are happy accidents. Some are just accidents. Without fail, though, they are always somehow still fun.

We did not dwell on the flavour of the beer. The beer, this time, was not the point. We loved the idea that a brewery in Alberta would do this for bands. That it would welcome a group of musicians into the brewhouse and say, "Do your worst," even if they might. We loved that craft beer made all this happen. By that point, it was an almost unbearable amount of love. Even if it was by then a bit off-kilter, everything seemed right with the world, with Calgary, with Alberta.

"It's been really fun making our own beer," said Stevens, the singer, as the set ended with a crescendo. Once the music stopped, the room began to empty quickly. The lights came up, illuminating our nearly full pints on the table we stood beside. Feeling self-conscious, and wobbly, we called the ride-share for the last time. We'd done what we set out to do, though what exactly we had done none of us could say for sure. We were tired. Our states of mind were unclear. But we knew that, somewhere along the way, we got what we'd travelled here for. We'd seen Alberta from a different, crooked angle.

THE NEXT MORNING, Saturday, around 10:30, the ground was damp and the Calgary air cool and misty. Colin and I waited in the parking lot of our rundown hotel while Guy gathered his things in the room. Already, we were making plans to come back. At the rate that the industry is growing, I suggested, we could return for a night each year and probably never visit the same brewery twice. Maybe one day we'd be able to do the same thing at home, in Edmonton, which at the time had nothing quite like what we'd seen. Colin, dressed in an Annex T-shirt, shrugged and nodded. We now knew what was possible, and it gave us faith.

We were soon back on the Deerfoot, moving at a reasonable rate in the direction of home. About an hour up the highway, I began to see rolling fields, smaller but reminiscent of ones I'd passed a few weeks back with Kirk Zembal during our tour of the country-side, into which the roots of all taprooms reach. On the drive home from Lacombe that day, we spoke casually about the craft beer industry and our hopes for its future. Zembal, too, was full of faith. He even dared to compare Alberta to meccas of small-batch beer such as Portland, Oregon, and Boulder, Colorado. Maybe one day, as it is in places like that, he said, craft beer in Alberta will just be part of our lives. Special but not remarkable, like having the barber, butcher, and baker all in your neighbourhood.

Beer is not essential to life. It will not serve as the bedrock of the economy of a town, city, or province. But the thing that is essential, and the reason one might have high hopes for the industry, is the spirit that the industry represents. Craft beer as it exists today in Alberta was not born of necessity. It was born at the confluence of myriad social, economic, and cultural factors. What ultimately brought it into being was a great many people who saw an opportunity and chose to work together to take advantage of it. It was, and now is, bigger than beer. It is about the ways in which people choose to make their living and live their lives.

As the three of us sped steadily toward Edmonton, we'd been silent for probably more than an hour. Guy's highway gaze had remained virtually unbroken. Colin had been sleeping, waking occasionally and a bit eerily to sing along with a couple of lines of whatever was on the CD player before falling asleep again. Finally, out the window to my right I saw the familiar sight of a Welcome to Edmonton sign. Then we came up on Gateway Park, which seemed oddly unfamiliar for a moment. Suddenly I realized why.

For years, the park had been home to Leduc Number 1, the very oil derrick that, on February 13, 1947, sent a geyser of light crude fifteen metres into the air. Pictures of it—a flare of flame and smoke stretching up into the sky—look almost apocalyptic. But rather than an ending it was the beginning of something that changed the province forever and contributed to its identity as an energy powerhouse.

Today, as we zipped past and on toward the heart of the city, I noticed for the first time that the metal tower was gone. In November 2018, it had been taken down and shipped back to Leduc to stand at an oil and gas museum. Its days of being front and centre in the province's capital were over.

At first, the absence felt jarring. Compared to the hundreds of times I'd driven back into Edmonton, the park now seemed bereft, fundamentally changed. But as I turned to look back at the empty space, I could see how trees were already growing into the space left behind. The new was replacing the old. Maybe a little like the stacks of flats of mass-produced beer in a liquor store that were the only thing we saw in that space for decades, now being pushed aside to make room for homegrown product. It seemed perfectly natural, I thought, the way the space was already adapting.

ACKNOWLEDGEMENTS

THIS BOOK WOULD not exist were it not for the efforts and support of a great many people, and I hope that I do not miss any of them here.

From our first exchanges by email, Taryn Boyd and Renée Layberry at TouchWood Editions showed such enthusiasm for my project that they effectively created the illusion that I might pull it off. I owe them thanks for that, and to Taryn in particular for accommodating the way life undermines one's best-laid plans to meet a deadline, and also for agreeing to my request for Curtis Gillespie's editorial guidance and interventions. I am indebted to Curtis for years of mentorship, reassurance, and for never going easy on me, not to mention for all the effort he put into this book. Thank you, also, to Jennifer Cockrall-King for encouragement and practical advice very early in this process.

To my daughters, Linnea and Maeve, I apologize for having missed more than a few Saturday afternoons with you because of the work this required. Happily, that wrong was made into a right by my mother, Carolyn, who regularly welcomed them with unfaltering energy and love before sending me off to the library with a warning, rather than suggestion, not to hurry back. My father, Garry, earns credit for teasing the children throughout that time so as to inexplicably endear himself to them. Gratitude also goes to Dad for pointing out my run-on sentences and inconsistent use of commas.

The book had numerous other enablers. Among them are the many interviewees who gave generously of their time and insight. It was a privilege to hear their stories, and I hope I have done them justice. I'd be remiss in not mentioning and thanking Bryan Alary, the supervisor who was unquestioningly supportive of my efforts to find time to write this book while keeping a day job that I like and appreciate. Thank you also to Evan Baldwin, Colin Damo, and Guy Tebay for being willing accomplices along the way.

Most supportive of my efforts, however, and always to an extreme, was my wife, Leah. I can imagine it's bad enough being married to a writer of the occasional magazine article. But I cannot possibly imagine being married to one writing a book while also trying to be a decent husband and committed father. Thank you for your patience, understanding, and love.

ALBERTA BREWERIES

THIS LIST IS now almost certainly out of date. Don't blame me—blame all those enterprising men and women who keep opening craft breweries in Alberta. That said, it will be close. Still, consider cross-referencing with an online search or a good old-fashioned phone call if you're planning to make a special trip.

(Note: If a brewer did not have a stand-alone facility up and running *in Alberta* at the time of writing, it did not make it onto this list. Brewers are notoriously optimistic about opening dates, which tend to be delayed. Construction, permits, approvals, bureaucracy, money, and myriad other challenges have delayed some projects by years.)

CALGARY
Northwest
Brewsters Crowfoot
25 Crowfoot Terrace NW
403-208-BREW (2739)
brewsters.ca

Northeast
Caravel Craft Brewery
#12, 10221-15 St NE
caravelbrewery.com

Citizen Brewing Company
227-35 Ave NE
403-474-4677
citizenbrewingcompany.com

Common Crown Brewing Co.
943-28 St NE
587-356-4275
commoncrown.ca

Elite Brewing & Cidery
1319 Edmonton Trail
403-277-7099
elitebrewing.com

Heathen's Brewing
#7, 3300-14 Ave NE
heathensbrewing.ca

Minhas Micro Brewery
1314-44 Ave NE
403-695-3701
minhasbrewery.com

Railyard Brewing
#121, 10301-19 St NE
403-465-4831
railyardbrewing.ca

Rapid Ascent Brewery
#109, 10985-38 St NE
403-214-9813
rapidascentbrewing.com

Tool Shed Brewing Company
801-30 St NE
403-775-1749
toolshedbrewing.com

Zero Issue Brewing
4210-12 St NE
zeroissuebeer.com

Southwest
Brewsters 11th Ave
834-11 Ave SW
403-265-BREW (2739)
brewsters.ca

Inner City Brewing Company
820-11 Ave SW
587-880-8600
innercitybrewing.ca

Last Best Brewing & Distilling
607-11 Ave SW
587-353-7387
lastbestbrewing.com

Marda Loop Brewing
Company
3523-18 St SW
403-585-8087
mardaloopbrewing.com

Mill Street Brewery
219-17 Ave SW
403-454-6871
millstreetbrewery.com

Trolley 5 Restaurant & Brewery
728-17 Ave SW
403-454-3731
trolley5.com

Wild Rose Brewery
4580 Quesnay Wood Dr SW
403-727-5451
wildrosebrewery.com

Southeast
Annex Ale Project
4323-1 St SE
403-475-4412
annexales.com

Banded Peak Brewing
#119, 519-34 Ave SE
403-283-5133
bandedpeakbrewing.com

Big Rock Brewery
5555-76 Ave SE
403-720-3239
bigrockbeer.com

Boiling Oar Brewing Company
7930-51 St SE
403-903-5003
boilingoar.com

Born Colorado Brewing
#3, 414-36 Ave SE
403-247-0295
borncoloradobrewing.com

Bow River Brewing
#110, 5769-4 St SE
bowriverbrewing.com

Brewsters Foothills
5519-53 St SE
403-723-BREW (2739)
brewsters.ca

Brewsters Lake Bonavista
#176, 755 Lake Bonavista Dr SE
403-225-BREW (2739)
brewsters.ca

Brewsters McKenzie Towne
#100, 11 McKenzie Towne Ave SE
403-243-BREW (2739)
brewsters.ca

Cabin Brewing Company
505-36 Ave SE
403-244-3331
cabinbrewing.ca

Cold Garden Beverage Company
1100-11 St SE
403-764-2653
coldgarden.ca

The Dandy Brewing Company &
Tasting Room
2003-11 St SE
587-956-8836
thedandybrewingcompany.com

Eighty-Eight Brewing Company
#1070, 2600 Portland St SE
403-452-5880
eightyeightbrewing.ca

The Establishment
Brewing Company
4407-1 St SE
403-453-6212
establishmentbrewing.ca

Good Mood Brewery
#2123, 4416-64 Ave SE
403-452-1806
goodmoodbrewery.com

High Line Brewing
#113, 1318-9 Ave SE
587-786-BEER (2337)
highlinebrewing.com

Legend 7 Brewing
4025-9 St SE
587-355-3105
legend7brewing.com

New Level Brewing
#4140, 7005 Fairmount Dr SE
403-764-4359
newlevelbrewing.ca

Ol' Beautiful Brewing Co.
1103-12 St SE
403-453-2739
olbeautiful.com

The O.T. Brewing Company
1155-44 Ave SE
403-816-4575
otbrewingcompany.com

Outcast Brewing
607 Manitou Rd SE
403-477-5478
outcastbrewing.ca

Paddy's Barbecue & Brewery
3610 Burnsland Rd SE
403-651-7150
paddysbrewbecue.com

Prairie Dog Brewing
105D-58 Ave SE
403-407-2448
prairiedogbrewing.ca

Revival Brewcade
1217B-9 Ave SE
587-893-2337
revivalbrewcade.com

Village Brewery
5000-12A St SE
403-243-3327
villagebrewery.com

CENTRAL ALBERTA

Battle River Brewery
4720A-37 St
Camrose
battleriverbrewery.ca

Beerdmeister
1806-20 St
Didsbury
403-807-6476
beerdmeister.com

Blindman Brewing
Bay F, 3413-53 Ave
Lacombe
403-786-BEER (2337)
blindmanbrewing.com

Cowtown Brewing Company
1806-20 St
Didsbury
403-415-4151
cowtownbeer.ca

Dark Woods Brewery
4720-50 St
Innisfail
403-597-1546
darkwoodsbrewing.com

Field & Forge Brewing Co.
3775-61 Ave
Innisfail
403-227-7402
fieldandforge.ca

4th Meridian Brewing Co.
#6, 2626-50 Ave
Lloydminster
780-870-0819
4mbrewingco.com

Hamill Brothers Brewing
36501 Rge Rd 275
Penhold
403-506-3399
redshedmalting.ca

Hawk Tail Brewery
6311-52 St
Rimbey
403-843-3034
hawktailbrewery.com

Norsemen Brewing Co.
6505-48 Ave
Camrose
780-672-9171
norsemenbrewing.com

Olds College Brewery
4601-46 St
Olds
403-556-8293
ocbeer.ca

Prairie Brewing Company
Twp Rd 322
Kneehill County
403-443-8272
prairiebrewco.com

Ribstone Creek Brewery
4924-51 St
Edgerton
780-755-3008
ribstonecreekbrewery.ca

Siding 14 Brewing Company
3520-67 St
Ponoka
403-783-4001
siding14brewing.com

Snake Lake Brewing Company
26 Industrial Dr
Sylvan Lake
403-505-3349
snakelake.beer

Undercurrent Brewing Ltd.
5003 Lakeshore Dr
Sylvan Lake
undercurrentbrewing.ca

NORTHERN ALBERTA

Apex Predator Brewing
53527 Range Rd 181A
Yellowhead County
780-517-7008
apexpredatorbrewing.com

Cold Lake Brewing &
Distilling Co.
#4, 5109-51 Ave
Cold Lake
780-201-1611
coldlakebrewingdistilling.com

Dog Island Brewing
250 Caribou Trail SW
Slave Lake
780-666-4777
dogislandbrewing.com

Dunvegan Brewing Company
50080 AB Hwy 49
Rycroft
780-864-8110
dunveganbrewingco.
websitepro.hosting

Fat Unicorn Brewery
Plamondon
780-623-0280
fubrew.com

Folding Mountain Brewing
49321 Folding Mountain Village
Yellowhead County
780-817-6287
foldingmountain.com

GP Brewing Co.
8812-111A St
Grande Prairie
780-533-4677

Grain Bin Brewing Company
#101, 11707-97 Ave
Grande Prairie
780-830-0232
grainbinbeer.com

Heilan Beer House
9212-113 St
Fairview
heilanbeerhouse.ca

Jasper Brewing Company
624 Connaught Dr
Jasper
780-852-4111
jasperbrewingco.ca

Lakeland Brewing Company
4227-50 Ave
St Paul
780-614-9466
lakelandbrewing.wixsite.com

Peace River Brewing
#4, 9710-94 St
Peace River
peaceriverbrewing@gmail.com

EDMONTON AND AREA

Alley Kat Brewing Company
9929-60 Ave NW
780-436-8922
alleykatbeer.com

Analog Brewing Company
8620-53 Ave NW
587-990-9584
analogbrewing.ca

Bent Stick Brewing Co.
5416-136 Ave NW
bentstickbrewing.com

Blind Enthusiasm Brewing
Company
9570-76 Ave NW
blindenthusiasm.ca

Brewsters Castledowns
15327 Castledowns Rd
780-425-HOPS (4677)
brewsters.ca

Brewsters Century Park
2335-111 St
780-429-HOPS (4677)
brewsters.ca

Brewsters Meadowlark
15820-87 Ave
780-421-HOPS (4677)
brewsters.ca

Brewsters Oliver Square
11620-104 Ave NW
780-482-HOPS (4677)
brewsters.ca

Brewsters Summerside
1140-91 St SW
780-424-HOPS (4677)
brewsters.ca

Campio Brewing Co.
10257-105 St NW
587-635-1953
campiobrewingco.com

Endeavour Brewing Company
#4A, 215 Carnegie Dr
St. Albert
780-752-3777
endeavourbrewing.com

The Growlery Beer Co.
40 Airport Rd NW
587-497-7714
growlerybeer.com

The Monolith (by Blind
Enthusiasm)
9919-78 Ave NW
blindenthusiasm.ca

Odd Company Brewing
#105, 12021-102 Ave NW
587-590-9973
oddcompany.ca

Omen Brewing
9942-67 Ave NW
780-760-6636
omenbrewing.com

Polar Park Brewing Company
10416-80 Ave NW
780-242-2337
polarparkbrewing.com

Rural Routes Brewing
Company
4901-50 St
Leduc
587-274-2739
ruralroutesbrewing.ca

Sea Change Brewing Co.
9850-62 Ave NW
780-784-2996
seachangebeer.com

Situation Brewing
10308-81 Ave NW
780-705-1377
situationbeer.com

SYC Brewing Co.
11239-180 St NW
sycbrewing.com

Town Square Brewing Co.
2919 Ellwood Dr SW
780-244-0212
townsquarebrewing.com

Yellowhead Brewery
10229-105 St NW
780-423-3333
yellowheadbrewery.com

RED DEER

Belly Hop Brewing Company
8105 Edgar Industrial Dr
403-318-0853
bellyhopbrewing.com

Red Hart Brewing
#112, 488 McCoy Dr
Red Deer County
redhartbrewing.ca

Sawback Brewing Co.
7023 Johnstone Dr
587-823-8888
sawbackbrewery.ca

Something Brewing Company
6610-71 St
403-803-9543
somethingbeer.com

Troubled Monk Brewery
5551-45 St
403-348-2378
troubledmonk.com

SOUTHERN ALBERTA

Balzac Craft Brewing Co.
#306, 401 Coopers Blvd SW
Airdrie
587-254-1957
balzaccraftbrewing.com

Banff Ave Brewing Co.
#2, 110 Banff Ave
Banff
403-762-1003
banffavebrewingco.ca

Brauerei Fahr
123 Kennedy Dr SE
Turner Valley
fahr.ca

Brewsters Airdrie
#200, 3 Stonegate Dr NW
Airdrie
403-945-BREW (2739)
brewsters.ca

Canmore Brewing Company
1460 Railway Ave
Canmore
403-678-2337
canmorebrewing.com

Coulee Brew Co.
4085-2 Ave S
Lethbridge
403-394-2337
couleebrew.co

Fitzsimmons Brewing Company
#4, 220 East Lake Blvd NE
Airdrie
587-892-2739
fitzsimmonsbrewing.com

Grizzly Paw Brewing Company
622-8 St
Canmore
403-678-2487
thegrizzlypaw.com

Grizzly Paw Brewing Company
Brewery
310 Old Canmore Rd
Canmore
403-678-2487
thegrizzlypaw.com

Half Hitch Brewing Company
#1, 10 Griffin Industrial Pt
Cochrane
403-988-4214
halfhitchbrewing.ca

Hard Knox Brewery
445-1 Ave NE
Black Diamond
403-800-5603
hardknoxbrewery.com

Hell's Basement Brewery
#102, 552-18 St SW
Medicine Hat
403-487-0489
hellsbasement.com

High River Brewing
Company Ltd.
510-21 St SE
High River
403-649-2002
hrbrewco.com

Hub Town Brewing Co.
41 Elizabeth Street
Okotoks
403-826-6864
hubtownbrewing.com

Medicine Hat
Brewing Company
1366 Brier Park Dr NW
Medicine Hat
403-525-1260
medicinehatbrewingcompany.ca

Newell Brewing Company
#112, 328-7 St E
Brooks
403-793-2378
newellbrewing.ca

Oldman River Brewing
101 Breckenridge Ave
Lundbreck
403-751-0017
oldmanriverbrewing.com

Origin Malting & Brewing Co.
60 Spruce Park Dr
Strathmore
403-902-0868
originmalting.com

Piston Broke Brewing
350-9 St E
Brooks
403-501-4818
pistonbrokebrewing.com

Rocky View Brewing
Company
4 Willow Ln
Cochrane
403-710-8518
rockyviewbrewingco.ca

Sheepdog Brewing
112-105 Bow Meadows Cres
Canmore
403-679-4009
sheepdogbrewing.com

Spectrum Ale Works
3500-9 Ave N
Lethbridge
403-315-1833
spectrumaleworks.com

Stronghold Brewing Co.
230-24 St
Fort MacLeod
403-635-9381
strongholdbrewing.ca

[Theoretically]
Brewing Company
1263-2 Ave S
Lethbridge
403-715-5140
theorybrew.ca

Township 24 Brewery
100 Rainbow Rd
Chestermere
403-460-8696
township24.ca

Travois Ale Works
612-3 St SE
Medicine Hat
587-289-1000
travoisbeer.com

Valley Brewing
242-3rd Ave West
Drumheller
403-823-3823

SOURCES

pp. 117–18   Quotes from Russell Sobel come from *Demographics and Entrepreneurship: Mitigating the Effects of an Aging Population*, published by the Fraser Institute and edited by Steven Globerman and Jason Clemens, 2018.

pp. 132–33   Quote from Dan Clapson comes from "Dandy Brewing Company Serves Up Some of Alberta's Best Taproom Cuisine," published in *The Globe and Mail*, August 4, 2018.

p. 242   Quote from Barry Vacker comes from his essay "Honoring the 50th Anniversary of 2001: The Monolith and Hope for the Human Species," published on medium.com, May 9, 2017. medium.com/explosion-of-awareness/honoring-the-50th-anniversary-of-2001-the-monolith-and-hope-for-the-human-species-1704c93501a0.

p. 246   Quote from Arthur C. Clarke comes from his essay "Hazards of Prophecy: The Failure of Imagination," which first appeared in his 1962 book, *Profiles of the Future: An Enquiry into the Limits of the Possible*.

p. 264   Quote from Pico Iyer comes from a lecture he delivered at the Smithsonian Institution in Washington, D.C., in 1996, entitled "Why We Travel."

INDEX

ABOUT THE AUTHOR

SCOTT MESSENGER'S blog *One Year of Alberta Beer,* which was featured on CBC Radio and in other Edmonton media, led him to a deeper examination of the steadily booming craft beer industry in Alberta. His writing on a variety of subjects has appeared in *The Guardian, Eighteen Bridges, Canadian Geographic, Avenue,* and more. He lives in Edmonton with his wife and two daughters.